Astronomy in a nutshell

*The Chief Facts and Principles Explained in Popular
Language for the General Reader and for Schools*

Garrett P. Serviss

Originally Published by
G. P. Putnam's Sons
1912

Contents

PREFACE

How many thousands of educated people, trained in the best schools, or even graduates of the great universities, have made the confession: "I never got a grip on astronomy in my student days. They didn't make it either plain or interesting to me; and now I am sorry for it."

The purpose of the writer of this book is to supply the need of such persons, either in school, or at home, after school-days are ended. He does not address himself to special students of the subject—although they, too, may find the book useful at the beginning—but to that vast, intelligent public for whom astronomy is, more or less, a "mystical midland," from which, occasionally, fascinating news comes to their ears. The ordinary text-book is too overladen with technical details, and too summary in its treatment of the general subject, to catch and hold the attention of those who have no special preliminary interest in astronomy. The aim here is to tell all that really needs to be told, and no more, and to put it as perspicuously, compactly, and interestingly, as possible. For that reason the book is called a "nutshell."

The author has been sparing in the use of diagrams, because he believes that, in many cases, they have been over-pressed. There is a tendency to try to represent everything to the eye. This is well to a certain extent, but there is danger that by pursuing this method too far the power of mental comprehension will be weakened. After all, it is only by an intelligent use of the imagination that progress can be made in such a science as astronomy. The reader is urged to make a serious effort to understand what is said in the text, and to picture it in his mind's eye, before referring to the diagrams. After he has thus presented the subject to his imagination, he may refer to the illustrations, and correct with their aid any misapprehension. For this reason the cuts, with their descriptions, have been made independent of the regular text, although they are placed in their proper connections throughout the book.

G. P. S.

April, 1912.

Part I—The Celestial Sphere.

Part II—The Earth.

Part III—The Solar System.

Part IV—The Fixed Stars.

PART I. THE CELESTIAL SPHERE.

PART I. THE CELESTIAL SPHERE.

1. Definition of Astronomy. Astronomy has to do with the earth, sun, moon, planets, comets, meteors, stars, and nebulæ; in other words, with the universe, or "the aggregate of existing things." It is the most ancient of all sciences. The derivation of the name from two Greek words, *aster*, "star," and *nomos*, "law," indicates its nature. It deals with the *law of the stars*—the word "star" being understood, in its widest signification, as including every heavenly body of whatever kind. The earth itself is such a body. Since we happen to live on the earth, it becomes our standpoint in space, from which we look out at the others. But, if we lived on some other planet, we would see the earth as a distant body in the sky, just as we now see Jupiter or Mars.

Astronomy teaches us that everything in the universe, from the sun and the moon to the most remote star or the most extraordinary nebula, is related to the earth. All are made of similar elementary substances and all obey similar physical laws. The same substance which is a solid upon the earth may be a gas or a vapour in the sun, but that does not alter its essential nature. Iron appears in the sun in the form of a hot vapour, but fundamentally it is the same substance which exists on the earth as a hard, tough, and heavy metal. Its different states depend upon the temperature to which it is subjected. The earth is a cool body, while the sun is an intensely hot one; consequently iron is solid on the earth and vaporous in the sun, just as in winter water is solid ice on the surface of a pond and steamy vapour over the boiler in the kitchen. Even on the earth we can make iron liquid in a blast furnace, and with the still greater temperatures obtainable in a laboratory we can turn it into vapour, thus reducing it to something like the state in which it regularly exists in the sun.

This fact, that the entire universe is made up of similar substances, differing only in state according to the local circumstances affecting them, is the greatest thing that astronomy has to tell us. It may be regarded as the fundamental law of the stars.

2. The Situation of the Earth in theHeavens. One of the greatest triumphs of human intelligence is the discovery of the real place which the earth occupies in the universe. This discovery has been made in spite of the most deceptive appearances. If we accepted the sole evidence of our eyes, as men once did, we could only conclude that the earth was the centre of the universe. In the daytime we see the sun apparently moving through the sky from east to west, as if it were travelling in a circle round the earth, overhead by day and underfoot at night. In the night-time, we see the stars apparently travelling round the earth in the same way as the sun. The fact is, that all of them are virtually motionless with regard to the earth, and their apparent movements through the sky are produced by the earth's rotation on its axis. The earth turns round on itself once every twenty-four hours, like a spinning ball. Imagine a fly on a rotating school globe; the whole room would appear to the fly to be revolving round it as the heavens appear to revolve round the earth. It would have to be a very intelligent insect to correct the deceptive evidence of its eyes.

The actual facts, revealed by many centuries of observation and reasoning, are that the earth is a rotating globe, turning once on its axis every twenty-four hours and revolving once round the sun every three hundred and sixty-five days. The sun is also a globe, 1,300,000 times larger than the earth, but so hot that it glows with intense brilliance, while the substances of which it consists are kept in a gaseous or vaporous state. Besides the earth there are seven other principal globes, or planets, which revolve round the sun, at various distances and in various periods, and, in addition to these, there are hundreds of smaller bodies, called asteroids or small planets. There are also many singular bodies called comets, and swarms of still smaller ones called meteors, which likewise revolve round the sun.

The earth and the other bodies of which we have just spoken are not only cooler than the sun, but most of them are in a solid state and do not shine with light of their own. The sun furnishes both heat and light to the smaller and cooler bodies revolving round it. In fact, the sun is simply a *star*, resembling the thousands of other stars which surround us in the sky, and its apparent superiority to them is due only to the fact that it is relatively near-by while they are far away. It is probable that all, or most, of the stars also have planets, comets, and meteors revolving round them, but invisible owing to their immense distance.

The "paths" in which the earth and the other planets and bodies travel in their revolutions round the sun are called their orbits. These orbits are all elliptical in shape, but those of the earth and the other large planets are not very different from circles. Some of the asteroids, and all of the comets, however, travel in elliptical orbits of considerable eccentricity, *i. e.*, which differ markedly from circles. The orbit of the earth differs so slightly from a circle that the eccentricity amounts to only about one-sixtieth. The distance of the earth from the sun being, on the average, 93,000,000 miles, the eccentricity of its orbit causes it to approach to within about 91,500,000 miles in winter (of the northern hemisphere) and to recede to about 94,500,000 miles in summer. The point in its orbit where the earth is nearest the sun is called perihelion, and the point where it is farthest from the sun, aphelion. The earth is at perihelion about Jan. 1, and at aphelion about July 4.

Now, in order to make a general picture in the mind of the earth's situation, let the reader suppose himself to be placed out in space as far from the sun as from the other stars. Then, if he could see it, he would observe the earth as a little speck, shining like a mote in the sunlight, and circling in its orbit close around the sun. The universe would appear to him to be somewhat like an immense spherical room filled with scattered electric-light bulbs, suspended above, below, and all around him, each of these bulbs representing a sun, and if there were minute insects flying around each light, these insects would represent the planets belonging to the various suns. One of the glowing bulbs among the multitude would stand for our sun, and one of the insects circling round it would be the earth.

Photograph of the South Polar Region of the Moon

Made by G. W. Ritchey with the forty-inch refractor of the Yerkes Observatory.

We have already remarked that the rotation of the earth on its axis causes all the other heavenly bodies to *appear* to revolve round it once every twenty-four hours, and we must now add that the earth's revolution round the sun causes the same bodies to *appear* to make another, slower revolution round it once every year. This introduces a complication of apparent motions which it is the business of astronomy to deal with, and which we shall endeavour to explain.

3. The Horizon, the Zenith, and theMeridian. First, let us consider what is the ordinary appearance of the sky. When we go out of doors on a clear night we see the heavens in the shape of a great dome arched above us and filled with stars. What we thus see is one half of the spherical shell of the heavens which surrounds us on all sides, the earth being apparently placed at its centre. The other half is concealed from our sight behind, or below, the earth. This spherical shell, of which only one half is visible to us at a time, is called the celestial sphere. Now, the surface of the earth seems to us (for this is another of the deceptive appearances which astronomy has to correct) to be a vast flat expanse, whose level is broken by hills and mountains, and the visible half of the celestial sphere seems to bend down on all sides and to rest upon the earth in a circle which extends all around us. This circle, where the heavens and the earth appear to meet, is called the horizon. As we ordinarily see it, the horizon appears irregular and broken on account of the unevenness of the earth's surface, but if we are at sea, or in the midst of a great level prairie, the horizon appears as a smooth circle, everywhere equally distant from the eye. This circle is called the sensible horizon. But there is another, ideal, horizon, used in astronomy, which is called the rational horizon. It is of the utmost importance that we should clearly understand what is meant by the rational horizon, and for this purpose we must consider another fact concerning the dome of the sky.

We now turn our attention to the centre of that dome, which, of course, is the point directly overhead. This point, which is of primary importance, is called the zenith. The position of the zenith is indicated by the direction of a plumb-line. If we imagine a plumb-line to be suspended from the centre of the sky overhead, and to pass into the earth at our feet, it would run through the centre of the earth, and, if it were continued onward in the same direction, it would, after emerging from the other side of the earth, reach the centre of the invisible half of the sky-dome at a point diametrically opposite to the zenith. This central point of the invisible half of the celestial sphere, lying under our feet, is called the nadir.

Keeping in mind the definitions of zenith and nadir that have just been given, we are in a position to understand what the rational horizon is. It is a great circle whose plane cuts through the centre of the earth, and which is situated exactly half-way between the zenith and the nadir. This plane is necessarily perpendicular, or at right angles, to the plumb-line joining the zenith and the nadir. In other words, the rational horizon divides the celestial sphere into two precisely equal halves, an upper and a lower half. In a hilly or mountainous country the sensible or visible horizon differs widely from the rational, or true horizon, but at sea the two are nearly identical. This arises from the fact, that the earth is so excessively small in comparison with the distances of most of the heavenly bodies that it may be regarded as a mere point in the midst of the celestial sphere.

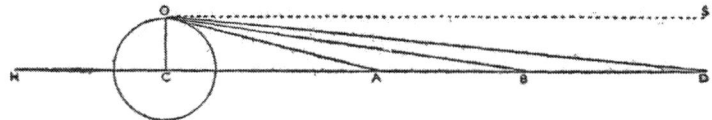

Fig. 1. The Rational and the Sensible Horizon.

Let C be the earth's centre, O the place of the observer, and H D the rational horizon passing through the centre of the earth. For an object situated near the earth, as at A, the sensible horizon makes a large angle with the rational horizon. If the object is farther away, as at B, the angle becomes less; and still less, again, if the object is at D. It is evident that if the object be immensely distant, like a star, the sensible horizon O S will be practically parallel with the rational horizon, and will blend with it, because the radius, or semi-diameter, of the earth, O C, is virtually nothing in comparison with the distance of the star.

Besides the horizon and the zenith there is one other thing of fundamental importance which we must learn about before proceeding further,—the meridian. The meridian is an imaginary line, or semicircle, beginning at the north point on the horizon, running up through the zenith, and then curving down to the south point. It thus divides the visible sky into two exactly equal halves, an eastern and a western half. In the ordinary affairs of life we usually think only of that part of the meridian which extends from the zenith to the south point on the horizon (which is sometimes called the "noon-line" because the sun crosses it at noon), but in astronomy the northern half of the meridian is as important as the southern.

4. Altitude and Azimuth. Now, suppose that we wish to indicate the location of a star, or other object, in the sky. To do so, we must have some fixed basis of reference, and such a basis is furnished by the horizon and the zenith. If we tried to describe the position of a star, the most natural thing would be, first, to estimate, or measure, its height above the horizon, and, second, to indicate the direction in which it was situated with regard to the points of the compass. These two measures, if they were accurately made, would enable another person to find the star in the sky. And this is precisely what is done in astronomy. The height above the horizon is called altitude, and the bearing with reference to the points of the compass is called azimuth. Together these are known as co-ordinates. In order to systematise this method of measuring the location of a star, the astronomer uses imaginary circles drawn on the celestial sphere. The horizon and the meridian are two of these circles. In addition to these, other imaginary circles are drawn parallel to the horizon and becoming smaller and smaller until the uppermost one may run close round the zenith, which is the common centre of the entire set. These are called altitude circles, because each one throughout its whole extent is at an unvarying height, or altitude, above the horizon. Such circles may be drawn anywhere we please, so as to pass through any chosen star or stars. If two stars in different quarters of the sky are found to lie on the same circle, then we know that both have the same altitude.

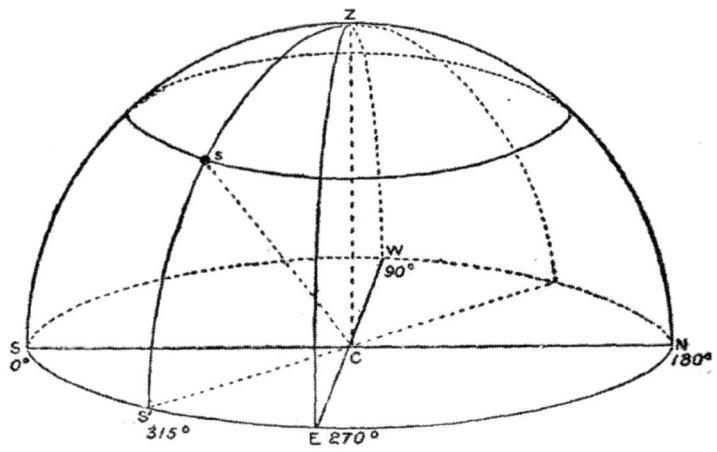

Fig. 2. Altitude and Azimuth.

C is the place of the observer.

N C S, a north-and-south line drawn in the plane of the horizon.

E C W, an east-and-west line in the plane of the horizon.

N E S W, the circle of the horizon.

Z, the observer's zenith.

N Z S, vertically above N C S, the meridian.

E Z W, the prime vertical.

Z s s , part of a vertical circle drawn through the star s.

The circle through s parallel to the horizon is an altitude circle.

The angle s C s , or the arc s s, represents the star's altitude.

The angle s C Z, or the arc Z s, is the star's zenith distance.

To find the azimuth, the angular distance round the horizon from S (0°), through W, N, E, to the point where the star's vertical circle meets the horizon, is measured. In this case it is 315°. But if we measured it eastward from the south point it would be—45°.

Then another set of circles is drawn perpendicular to the horizon, and all intersecting at the zenith and the nadir. These are called vertical circles, from the fact that they are upright to the horizon. That one of the vertical circles which cuts the horizon at the north-and-south points coincides with the meridian, which we have already described. The vertical circle at right angles to the meridian is called the prime vertical. It cuts the horizon at the east and west points, dividing the visible sky into a northern and a southern half. Like the altitude circles, vertical circles may be drawn anywhere we please so as to pass through a star in any quarter of the sky—but the meridian and the prime vertical are fixed.

With the two sets of circles that have just been described, it is possible to indicate accurately the location of any heavenly body, at any particular moment. Its altitude is ascertained by measuring, along the vertical circle passing through it, its distance from the horizon. (Sometimes it is convenient to measure, instead of the altitude of a star, its zenith distance, which is also reckoned on the vertical circle.)

To ascertain the azimuth, we must first choose a point of beginning on the horizon. Any of the cardinal points, *i.e.*, east, west, north, or south, may be employed for this purpose, but in astronomy it is customary to use only the south point, and to carry the measure westward all round the circle of the horizon, and so back to the point of beginning in the south. This involves circular, or degree, measure, to which a few words must now be devoted.

Every circle, no matter how large or how small, is divided into 360 equal parts, called degrees, usually indicated by the sign (°); each degree is subdivided into 60 equal parts called minutes, indicated by the sign ( ); and each minute is subdivided into 60 equal parts called seconds, indicated by the sign ( ). Thus there are 360°, or 21,600 , or 1,296,000 in every complete circle. The actual length of a degree in inches, yards, or miles, depends upon the size of the circle, but no circle ever has more than 360°, and a degree of any particular circle is precisely equal to any other degree of that same circle. Thus, if a circle is 360 miles in circumference, every one of its degrees will be one mile long. In mathematics, a degree usually means not a distance measured along the circumference of a circle, but an angle formed at the centre of the circle between two lines called radii (radius in the singular), which lines, where they intersect the circumference, are separated by a distance equal to one 360th of the entire circle. But, for ordinary purposes, it is simpler to think of a degree as an arc equal in length to one 360th of the circle. Now, since the horizon, and the other imaginary lines drawn in the sky, are all circles, it is evident that the principle of circular measure may be applied to them, and indeed must be so applied in order that they shall be of use to us in indicating the position of a star.

To return, then, to the measurement of the azimuth of a star. Since the south point is the place of beginning, we mark it 0°, and we divide the circle of the horizon into 360°, counting round westward. Suppose we see a star somewhere in the south-western quarter of the sky; then the point where the vertical circle passing through that star intersects the horizon will indicate its azimuth. Suppose that this point is found to be 25° west of south; then 25° will be the star's azimuth. Suppose it is 90°; then the azimuth is 90°, and the star must be on the prime vertical in the west, because west, being one quarter of the way round the horizon from south, is 90° in angular distance from the south point. Suppose the azimuth is 180°; then the star must be on the meridian north of the zenith, because north is exactly half-way, or 180°

round the horizon from the south point. Suppose the azimuth is 270°; then the star must be on the prime vertical in the east, because east is 270°, or three quarters of the way round from the south point. If the star is on the meridian in the south its azimuth may be called either 0° or 360°, because on any graduated circle the mark indicating 360° coincides in position with 0°, that being at the same time the point of beginning and the point of ending.

The same system of angular measure is applied in ascertaining a star's altitude. Since the horizon is half-way between the zenith and the nadir it must be just 90° from either. If a star is in the zenith, then its altitude is 90°, and if it is below the zenith its altitude lies somewhere between 0° and 90°. In any case it cannot be less than 0° nor more than 90°. Having measured the altitude and the azimuth we have the two co-ordinates which are needed to indicate accurately the place of a star in the sky. But, as we shall see in a moment, other co-ordinates beside altitude and azimuth are needed for a complete description of the places of the stars on the celestial sphere. Owing to the apparent revolution of the heavens round the earth, the altitudes and azimuths of the celestial bodies are continually changing. We shall now study the causes of these changes.

5. The Apparent Motion of the Heavens. We have likened the earth to a rotating school globe. As such a globe turns, any particular spot on it is presented in succession toward the various sides of the room. In precisely the same way any spot on the earth is turned by its rotation successively toward various parts of the surrounding sky. To understand the effect of this, a little patient watching of the actual heavens will be required, but this has the charm of all out-of-doors observation of nature, and it will be found of fascinating interest as the facts begin to unfold themselves.

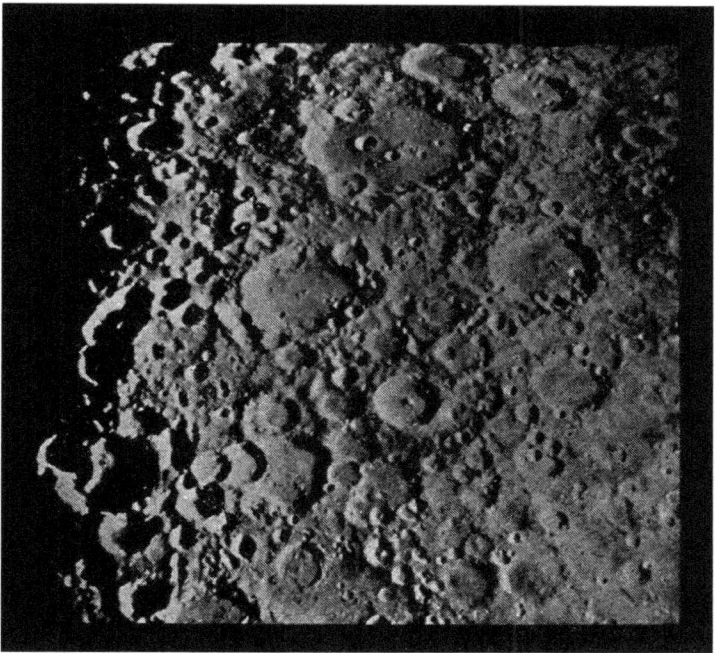

The Moon Near the "Crater" Tycho

Photographed at the Lick Observatory under the direction of E. S. Holden.

Tycho is the regular oval depression a little below the centre of the view. The vast depression, 140 miles across, with a row of smaller craters within, below the centre of the view at the top, is Clavius. The photograph was made when sundown was approaching on that part of the moon. Observe the jagged line of advancing night lying across the rugged surface on the western (left-hand) side.

It is best to begin by finding the North Star, or pole star. If you are living not far from latitude 40° north, which is the median latitude of the United States, you must, after determining as closely as you can the situation of the north point, look upward along the meridian in the north until your eyes are directed to a point about 40° above the horizon. Forty degrees is somewhat less than half-way from the horizon to the zenith, which, as we have seen, are separated by an arc of 90°. At that point you will notice a lone star of what astronomers call the second magnitude. This is the celebrated North Star. It is the most useful to man of all the stars, except the sun, and it differs from all the others in a way presently to be explained. But first it is essential that you should make no mistake in identifying it. There are certain landmarks in the sky which make such identification certain. In the first place, it is always so close to the meridian in the north, that by naked-eye observation you would probably never suspect that it was not exactly on the meridian.

Then, its altitude is always equal, or very nearly equal, to the latitude of the place where you happen to be on the earth, so that if you know your latitude you know how high to carry your eye above the northern horizon. If you are in latitude 50°, the star will be at 50° altitude, and if your latitude is 30°, the altitude of the star will be 30°. Next you will notice that the North Star is situated at the end of the handle of a kind of dipper-shaped figure formed by stars, the handle being bent the wrong way. All of the stars forming this "dipper" are faint, except the two which are farthest from the North Star, in the outer edge of the bowl, one of which is about as bright as the North Star itself. Again, if you carry your eye along the handle to the bowl, and then continue onward about as much farther, you will be led to another, larger, more conspicuous, and more perfect, dipper-shaped figure, which is in the famous constellation of Ursa Major, or the Great Bear. This striking figure is called the Great Dipper (known in England as The Wain). It contains seven conspicuous stars, all of which, with one exception, are equal in brightness to the North Star. Now, look particularly at the two stars which indicate the outer side of the bowl of this dipper, and you will find that if you draw an imaginary line through them toward the meridian in the north, it will lead your eye directly back to the North Star. These two significant stars are often called The Pointers. With their aid you can make sure that you have really found the North Star.

Having found it, begin by noting the various groups of stars, or constellations, in the northern part of the sky, and, as the night wears on, observe whether any change takes place in their position. To make our description more definite, we will suppose that the observations begin at nine o'clock P.M. about the 1st of July. At that hour and date, and from the middle latitudes of the United States, the Great Dipper is seen in a south-westerly direction from the North Star, with its handle pointing overhead. At the same time, on the opposite side of the North Star, and low in the north-east, appears the remarkable constellation of Cassiopeia, easily recognisable by a zigzag figure, roughly resembling the letter W, formed by its five principal stars. Fix the relative positions of these constellations in the memory, and an hour later, at 10 P.M., look at them again. You will find that they have moved, the Great Dipper sinking, while Cassiopeia rises. Make a third observation at 11 P.M., and you will perceive that the motion has continued, The Pointers having descended in the north-west, until they are on a level with the North Star, while Cassiopeia has risen to nearly the same level in the north-east. In the meantime, the North Star has remained apparently motionless in its original position. If you repeat the observation at midnight, you will find that the Great Dipper has descended so far that its centre is on a level with the North Star, and that Cassiopeia has proportionally risen in the north-east. It is just as if the two constellations were attached to the ends of a rod pivoted at the centre upon the North Star and twirling about it.

In the meantime you will have noticed that the figure of the "Little

Dipper," attached to the North Star, which had its bowl toward the zenith at 9 P.M., has swung round so that at midnight it is extended toward the south-west. Thus, you will perceive that the North Star is like a hub round which the heavens appear to turn, carrying the other stars with them.

To convince yourself that this motion is common to the stars in all parts of the sky, you should also watch the conduct of those which pass overhead, and those which are in the southern quarter of the heavens. For instance, at 9 P.M. (same date), you will see near the zenith a beautiful coronet which makes a striking appearance although all but one of its stars are relatively faint. This is the constellation Corona Borealis, or the Northern Crown. As the hours pass you will see the Crown swing slowly westward, descending gradually toward the horizon, and if you persevere in your observations until about 5 A.M., you will see it set in the north-west. The curve that it describes is concentric with those followed by the Great Dipper and Cassiopeia, but, being farther from the North Star than they are, and at a distance greater than the altitude of the North Star, it sinks below the horizon before it can arrive at a point directly underneath that star. Then take a star far in the south. At 9 o'clock, you will perceive the bright reddish star Antares, in the constellation Scorpio, rather low in the south and considerably east of the meridian. Hour after hour it will move westward, in a curve larger than that of the Northern Crown, but still concentric with it. A little before 10 P.M., it will cross the meridian, and between 1 and 2 A.M., it will sink beneath the horizon at a point south of west.

So, no matter in what part of the heavens you watch the stars, you will see not only that they move from east to west, but that this motion is performed in curves concentric round the North Star, which alone appears to maintain its place unchanged. Along the eastern horizon you will perceive stars continually rising; in the middle of the sky you will see others continually crossing the meridian—a majestic march of constellations,—and along the western horizon you will find still others continually setting. If you could watch the stars uninterruptedly throughout the twenty-four hours (if daylight did not hide them from sight during half that period), you would perceive that they go entirely round the celestial sphere, or rather that it goes round with them, and that at the end of twenty-four hours they return to their original positions. But you can do this just as well by looking at them on two successive nights, when you will find that at the same hour on the second night they are back again, practically in the places where you saw them on the first night.

Of course, what happens on a July night happens on any other night of the year. We have taken a particular date merely in order to make the description clearer. It is only necessary to find the North Star, the Great Dipper, and Cassiopeia, and you can observe the apparent revolution of the heavens at any time of the year. These constellations, being so near the North Star that they never go entirely below the horizon in middle northern latitudes, are always visible on one side or another of the North Star.

Now, call upon your imagination to deal with what you have been observing, and you will have no difficulty in explaining what all this apparent motion of the stars means. You already know that the heavens form a sphere surrounding the earth. You have simply to suppose the North Star to be situated at, or close to, the north end, or north pole, of an imaginary axle, or axis, round which the celestial sphere seems to turn, and instantly the whole series of phenomena will fall into order, and the explanation will stare you in the face. That explanation is that the motion of all the stars in concentric circles round the North Star is due to an apparent revolution of the whole celestial sphere, like a huge hollow ball, about an axis, the position of one of whose poles is graphically indicated in the sky by the North Star. The circles in which the stars seem to move are perpendicular to this axis, and inclined to the horizon at an angle depending upon the altitude of the North Star at the place on the earth where the observations are made.

Another important fact demands our attention, although the thoughtful reader will already have guessed it—the north pole of the celestial sphere, whose position in the sky is closely indicated by the North Star, is situated directly over the north pole of the earth. This follows from the fact that the apparent revolution of the celestial sphere is due to the real rotation of the earth. You can see that the two poles, that of the earth and that of the heavens, must necessarily coincide, by taking a school globe and imagining that you are an intelligent little being dwelling on its surface. As the globe turned on its axis you would see the walls of the room revolving round you, and the poles of the apparent axis round which the room turned, would, evidently, be directly over the corresponding poles of the globe itself. Another thing which you could make clear by this experiment is that, as the poles of the celestial sphere are over the earth's poles, so the celestial equator, or equator of the heavens, must be directly over the equator of the earth.

We can determine the location of the poles of the heavens by watching the revolution of the stars around them, and we can fix the position of the circle of the equator of the heavens, by drawing an imaginary line round the celestial sphere, half-way between the two poles. We have spoken specifically only of the north pole, but, of course, there is a corresponding south pole situated over the south pole of the earth, but whose position is invisible from the northern hemisphere. It happens that the place of the south celestial pole is not indicated to the eye, like that of the northern, by a conspicuous star.

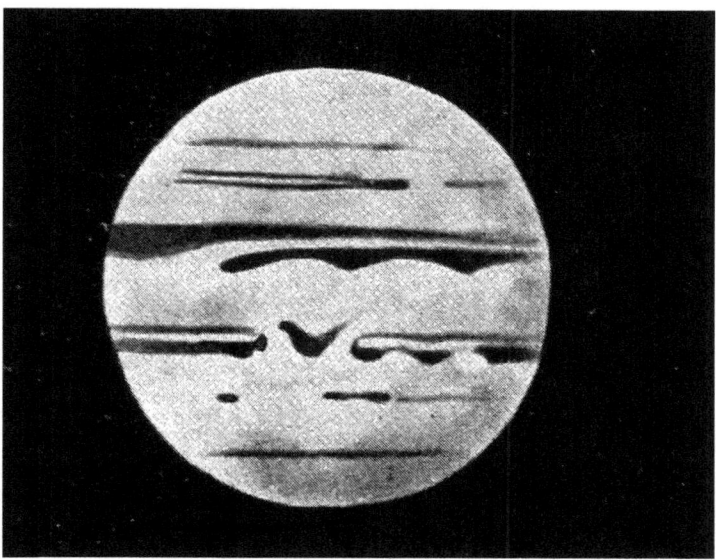

Drawing of Jupiter

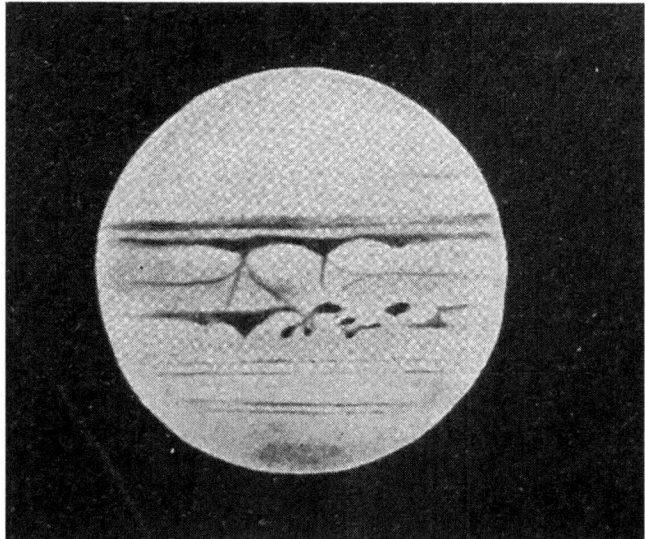

Drawing of Jupiter

Note the change of details between the two drawings, made at different times. Similar changes are continually occurring.

6. Locating the Stars on the CelestialSphere. Having found the poles and the equator of the celestial sphere, we begin to see how it is possible to make a map, or globe, of the heavens just as we do of the earth, on which the objects that they contain may be represented in their proper positions. When we wish to describe the location of an object on the earth, a city for instance, we have to refer to a system of imaginary circles, drawn round the earth and based upon the equator and the poles. These circles enable us to fix the place of any point on the earth with accuracy. One set of circles called parallels of latitude are drawn east and west round the globe parallel to the equator, and becoming smaller and smaller until the smallest runs close round their common central point, which is one of the poles. Each pole of the earth is the centre of such a set of circles all parallel to the equator. Since each circle is unvarying in its distance from the equator, all places which are situated anywhere on that circle have the same latitude, or distance from the equator, either north or south.

But to know the latitude of any place on the earth is not sufficient; we must also know what is called its longitude, or its angular distance east or west of some chosen point on the equator. This knowledge is obtained with the aid of another set of circles drawn north-and-south round the earth, and all meeting and crossing at the poles. These are called meridians of longitude. In order to make use of them we must, as already intimated, select some particular meridian whose crossing point on the equator will serve as a place of beginning. By common consent of the civilised world, the meridian which passes through the observatory at Greenwich, near London, has been chosen for this purpose. It is, like all the meridians, perpendicular to the equator, and it is called the prime meridian of the earth.

In locating any place on the earth, then, we ascertain by means of the parallel of latitude passing through it how far in degrees, it is north or south of the equator, and by means of its meridian of longitude how far it is east or west of the prime meridian, or meridian of Greenwich. These two things being known, we have the exact location of the place on the earth. Let us now see how a similar system is applied to ascertain the location of a heavenly body on the celestial sphere.

We have observed that the poles of the heavens correspond in position, or direction, with those of the earth, and that the equator of the heavens runs round the sky directly over the earth's equator. It follows that we can divide the celestial sphere just as we do the surface of the earth by means of parallels and meridians, corresponding to the similar circles of the earth. On the earth, distance from the equator is called latitude, and distance from the prime meridian, longitude. In the heavens, distance from the equator is called declination, and distance from the prime meridian, right ascension; but they are essentially the same things as latitude and longitude, and are measured virtually in the same way. In place of parallels of latitude, we have on the celestial sphere circles drawn parallel to the equator and centring about the celestial poles, which are called parallels of declination, and in place of meridians of longitude, we have circles perpendicular to the equator, and drawn through the celestial poles, which are called hour circles. The origin of this name will be explained in a moment. For the present it is only necessary to fix firmly in the mind the fact that these two systems of circles, one on the earth and the other in the heavens, are fundamentally identical.

Just as on the earth geographers have chosen a particular place, viz. Greenwich, to fix the location of the terrestrial prime meridian, so astronomers have agreed upon a particular point in the heavens which serves to determine the location of the celestial prime meridian. This point, which lies on the celestial equator, is known as the vernal equinox. We shall explain its origin after having indicated its use. The hour circle which passes through the vernal equinox is the prime meridian of the heavens, and the vernal equinox itself is sometimes called the "Greenwich of the Sky."

If, now, we wish to ascertain the exact location of a star on the celestial sphere, as we would that of New York, London, or Paris, on the earth, we measure along the hour circle running through it, its declination, or distance from the celestial equator, and then, along its parallel of declination, we measure its right ascension, or distance from the vernal equinox. Having these two co-ordinates, we possess all that is necessary to enable us to describe the position of the star, so that someone else looking for it, may find it in the sky, as a navigator finds some lone island in the sea by knowing its latitude and longitude.

Declination, as we have seen, is simply another name for latitude, but right ascension, which corresponds to longitude, needs a little additional explanation. It differs from longitude, first, in that, instead of being reckoned both east and west from the prime meridian, it is reckoned only toward the east, the reckoning being continued uninterruptedly entirely round the circle of the equator; and, second, in that it is usually counted not in degrees, minutes, and seconds of arc, but in hours, minutes, and seconds of time. The reason for this is that, since the celestial sphere makes one complete revolution in twenty-four hours, it is convenient to divide the circuit into twenty-four equal parts, each corresponding to the distance through which the heavens appear to turn in one hour. This explains the origin of the term hour circles applied to the celestial meridians, which, by intersecting the equator, divide it into twenty-four equal parts, each part corresponding to an hour of time. In expressing right ascension in time, the Roman numerals—I, II, III, IV, V, VI, VII, VIII, IX, X, XI, XII, XIII, XIV, XV, XVI, XVII, XVIII, XIX, XX, XXI, XXII, XXIII, XXIV—are employed for the hours, and the letters *m* and *s* respectively for the minutes and seconds. Since there are 360° in every circle, it is plain that one hour of right ascension corresponds to 15°. So, too, one minute of right ascension corresponds to 15 , and one second to 15 . It will be found useful to memorise these relations.

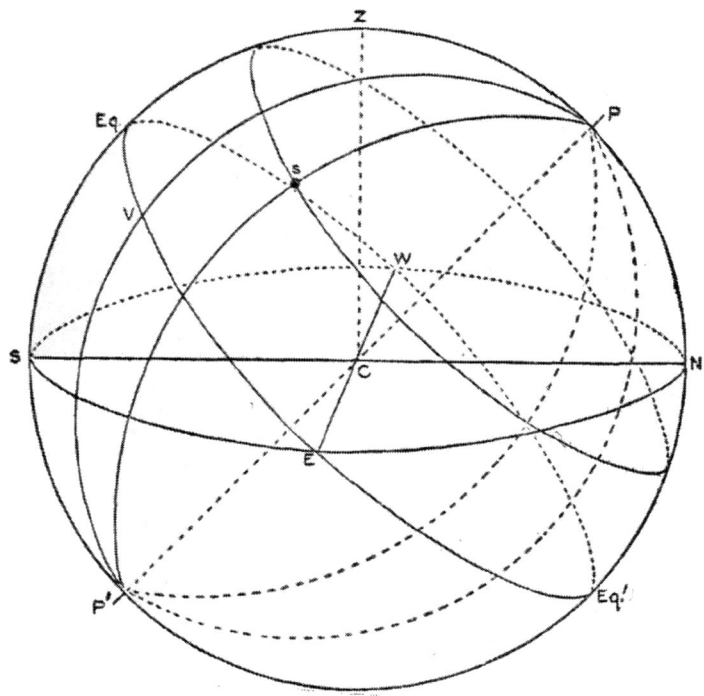

Fig. 3. Right Ascension and Declination.

The plane of the horizon, with the north, south, east, and west points, and the zenith, are represented as in Fig. 2.

P and P are the poles of the celestial sphere, the dotted line connecting them representing the direction of the axis, both of celestial sphere and the earth.

The circle Eq Eq is the equator.

V is the vernal equinox, or the point on the equator whence right ascension is reckoned round toward the east.

The circle passing through s, parallel to the equator, is a declination circle.

The circle P s P is the hour circle of the star s.

The arc of this hour circle contained between s and the point where it meets the equator is the star's declination. Its right ascension is measured by the arc of the equator contained between V and the point where its hour circle meets the equator, or by the angle V P s.

The hour circle P V P , passing through the vernal equinox, is the equinoctial colure. When this has moved up to coincidence with the meridian, N Z S, it will be astronomical noon.

7. Effects Produced by Changing the Observer's Place on the Earth. The reader will recall that in Sect. 4 we described another system of circles for determining the places of stars, a system based on the horizon and the zenith. This horizon-zenith system takes no account of the changes produced by the apparent motion of the heavens, and consequently it is not applicable to determining the absolute positions of the stars on the celestial sphere. It simply shows their positions in the visible half of the sky, as seen at some particular time from some definite point on the earth. In order to show the changing relations of this system to that which we have just been describing, let us consider the effects produced by shifting our place of observation on the earth. Since the zenith is the point overhead and the nadir the point underfoot, and the horizon is a great circle drawn exactly half-way between the zenith and the nadir, it is evident, upon a moment's consideration, that every place on the earth has its own zenith and its own horizon. It is also clear that every place must have its own meridian, because the meridian is a north-and-south line running directly over the observer's head. You can see how this is, if you reflect that for an observer situated on the other side of the earth what is overhead for you will be underfoot for him, and *vice versa*. Thus the direction of our zenith is the direction of the nadir for our antipodes, and the direction of their zenith is the direction of our nadir. They see the half of the sky which is invisible to us, and we the half which is invisible to them.

Now, suppose that we should go to the north pole. The celestial north pole would then be in our zenith, and the equator would correspond with the horizon. Thus, for an observer at the north pole the two systems of circles that we have described would fall into coincidence. The zenith would correspond with the pole of the heavens; the horizon would correspond with the celestial equator; the vertical circles would correspond with the hour circles; and the altitude circles would correspond with the circles of declination. The North Star, being close to the pole of the heavens, would appear directly overhead. Being at the zenith, its altitude would be 90° (see Sect. 4). Peary, if he had visited the pole during the polar night, would have seen the North Star overhead, and it would have enabled him with relatively little trouble to determine his exact place on the earth, or, in other words, the exact location of the north terrestrial pole. With the pole star in the zenith, it is evident that the other stars would be seen revolving round it in circles parallel to the horizon. All the stars situated north of the celestial equator would be simultaneously and continuously visible. None of them would either rise or set, but all, in the course of twenty-four hours, would appear to make a complete circuit horizontally round the sky. This polar presentation of the celestial sphere is called the parallel sphere, because the stars appear to move parallel to the horizon. No man has yet beheld the nocturnal phenomena of the parallel sphere, but if in the future some explorer should visit one of the earth's poles during the polar night, he would behold the spectacle in all its strange splendour.

Jupiter

Photographed at the Lick Observatory.

Observe on the left the Great Red Spot, which first appeared in 1878.

Next, suppose that you are somewhere on the earth's equator. Since the equator is everywhere 90°, or one quarter of a circle, from each pole, it is evident that looking at the sky from the equator you would see the two poles (if there was anything to mark their places) lying on the horizon one exactly in the north and the other exactly in the south. The celestial equator would correspond with the prime vertical, passing east and west directly over your head, and all the stars would rise and set perpendicularly to the horizon, each describing a semicircle in the sky in the course of twelve hours. During the other twelve hours, the same stars would be below the horizon. Stars situated near either of the poles would describe little semi-circles near the north or the south point; those farther away would describe larger semi-circles; those close to the celestial equator would describe semi-circles passing overhead. But all, no matter where situated, would describe their visible courses in the same period of time. This equatorial presentation of the celestial sphere is called the right sphere, because the stars rise and set at a right angle to the plane of the horizon. Comparing it with the system of circles on which right ascension and declination are based, we see that, as the prime vertical corresponds with the celestial equator, so the horizon must represent an hour circle. The meridian also represents an hour circle. It may require a little thought to make this clear, but it will be a good exercise.

Finally, if you are somewhere between the equator and one of the poles, which is the actual situation of the vast majority of mankind, you see either the north or the south pole of the heavens elevated to an altitude corresponding with your latitude, and the stars apparently revolving round it in circles inclined to the horizon at an angle depending upon the latitude. The nearer you are to the equator, the steeper this angle will be. This ordinary presentation of the celestial sphere is called the oblique sphere. Its horizon does not correspond with either the equator or the prime vertical, and its zenith and nadir lie at points situated between the celestial poles and the celestial equator.

8. The Astronomical Clock and the Ecliptic. It will be remembered that the meridian of any place on the earth is a straight north and south line running through the zenith and perpendicular to the horizon. More strictly speaking, the meridian is a circle passing from north to south directly overhead and corresponding exactly with the meridian of longitude of the place of observation. Now, let us consider the hour circles on the celestial sphere. They are drawn in the same way as the meridians on the earth. But the celestial sphere appears to revolve round the earth, and as it does so it must carry the hour circles with it, since they are fixed in position upon its surface. Fix your attention upon the first of these hour circles, *i. e.*, the one which runs through the vernal equinox. Its right ascension is called 0 hours, because it is the starting point. Suppose that at some time we find the vernal equinox exactly in the south; then the 0 hour circle, or the prime meridian of the heavens, will, at that instant, coincide with the meridian of the place of observation. But one hour later, in consequence of the motion of the heavens, the vernal equinox, together with the circle of 0 hours, will be 15°, or one hour of right ascension, west of the meridian, and the hour circle marked I will have come up to, and for an instant will be blended with, the meridian. An hour later still, the circle of II hours right ascension will have taken its place on the meridian, while the vernal equinox and the circle of 0 hours will be II hours, or 30°, west of the meridian. And so on, throughout the entire circuit of the sky.

What has just been said makes it evident that the apparent motion of the heavens resembles the movement of a clock, the vernal equinox, or the circle of 0 hours, serving as a hand, or pointer, on the dial. Astronomers use it in exactly that way, for astronomical clocks are made with dials divided into twenty-four hour spaces, and the time reckoning runs continuously from 0 hours to XXIV hours. The "astronomical day" begins when the vernal equinox is on the meridian. At that instant the hands of the astronomical clock mark 0 hours, 0 minutes, 0 seconds. Thus the clock follows the motion of the heavens, and the astronomer can tell by simply glancing at the dial, and without looking at the sky, where the vernal equinox is, and what is the right ascension of any body which may at that moment be on the meridian.

We must now explain a little more fully what the vernal equinox is, and why it has been chosen as the "Greenwich of the Sky." Its position is not marked by any star, but is determined by means of the intersecting circles that we have described. There is one other such circle, that we have not yet mentioned, which bears a peculiar relation to the vernal equinox. This is the ecliptic. Just as the daily, or diurnal, rotation of the earth on its axis causes the whole celestial sphere to appear to make one revolution every day, so the yearly or annual revolution of the earth in its orbit about the sun causes the sun to appear to make one revolution through the sky every year. As the earth moves onward in its orbit, the sun seems to move in the opposite direction. Inasmuch as there are 360° in a complete circle and 365 days in a year, the apparent motion of the sun amounts to nearly 1° per day, or 30° per month. In twelve months, then, the sun comes back again to the place in the sky which it occupied at the beginning of the year. Since the motion of the earth in its orbit is from west to east (the same as that of its rotation on its axis), it follows that the direction of the sun's apparent annual motion in the sky is from east to west (like its daily motion). Thus, while *infact* the earth pursues a path, or orbit, round the sun, the sun *seems* to pursue a path round the earth. This apparent path of the sun, projected against the background of the sky, is called the ecliptic. The name arises from the fact that eclipses only occur when the moon is in or near the plane of the sun's apparent path.

As the apparent motion of the sun round the ecliptic is caused by the real motion of the earth round the sun, we may regard the ecliptic as a circle marking the intersection of the plane of the earth's orbit with the celestial sphere. In other words, if we were situated on the sun instead of on the earth, we would see the earth travelling round the sky in the circle of the ecliptic. We must keep this fact, that the ecliptic indicates the plane of the earth's orbit, firmly in mind, in order to understand what follows.

The ecliptic is not coincident with the celestial equator, for the following reason: The axis of the earth's daily rotation is not parallel to, or does not point in the same direction as, the axis of its yearly revolution round the sun. As the axis of rotation is perpendicular to the equator, so the axis of the yearly revolution is perpendicular to the ecliptic, and since these two axes are inclined to one another, it results that the equator and the ecliptic must lie in different planes. The inclination of the plane of the ecliptic to that of the equator amounts to about 23½°.

As it is very important to have a clear conception of this subject, we may illustrate it in this way: Take a ball to represent the earth, and around it draw a circle to represent the equator. Then, through the centre of the ball, and at right angles to its equator, put a long pin to represent the axis. Set it afloat in a tub of water, weighting it so that it will be half submerged, and placing it in such a position that the pin will be not upright but inclined at a considerable angle from the vertical. Now, imagine that the sun is situated in the centre of the tub, and cause the ball to circle slowly round it, while maintaining the pin always in the same position. Then the surface of the water will represent the plane of the ecliptic, or plane of the earth's orbit, and you will see that, in consequence of the inclination of the pin, the plane of the equator does not coincide with that of the ecliptic (or the surface of the water), but is tipped with regard to it in such a manner that one half of the equator is below and the other half above it. Instead of actually trying this experiment, it will be a useful exercise of the imagination to represent it to the mind's eye just as if it were tried.

We have said that the inclination of the equator to the ecliptic amounts to 23½°, and this angle should be memorised. Now, since both the ecliptic and the equator are great circles of the celestial sphere, *i. e.*, circles whose planes cut through the centre of the sphere, they must intersect one another at two opposite points. In the experiment just described, these two points lie on opposite sides of the ball, where the equator cuts the level of the water. These points of intersection of the equator and the ecliptic on the celestial sphere are called the equinoxes, or equinoctial points, because when the sun appears at either of those points it is perpendicular over the equator, and when it is in that position day and night are of equal length all over the earth. (Equinox is from two Latin words meaning "equal night.")

Saturn
From a drawing by Trouvelot.

Saturn

Photographed at the Lick Observatory.

We shall have more to say about the equinoxes later, but for the present it is sufficient to remark that one of these points—that one where the sun is about the 21st of March, which is the beginning of astronomical spring— is the "Greenwich of the Sky," or the vernal equinox. The other, opposite, point is called the autumnal equinox, because the sun arrives there about the 23d of September, the beginning of astronomical autumn. The vernal equinox, as we have already seen, serves as a pointer on the dial of the sky. When it crosses the meridian of any place it is astronomical noon at that place. Its position in the sky is not marked by any particular star, but it is situated in the constellation Pisces, and lies exactly at the crossing point of the celestial equator and the ecliptic. The hour circle, running through this point, and through its opposite, the autumnal equinox, is the prime meridian of the heavens, called the equinoctial colure. The hour circle at right angles to the equinoctial colure, i. e., bearing to it the same relation that the prime vertical does to the meridian (see Sect. 4), is called the solstitial colure. This latter circle cuts the ecliptic at two opposite points, called the solstices, which lie half-way between the equinoxes. Since the ecliptic is inclined 23½° to the plane of the equator, and since the solstices lie half-way between the two crossing points of the ecliptic and the equator, it is evident that the solstices must be situated 23½° from the equator, one above and the other below, or one north and the other south. The northern one is called the summer solstice, because the sun arrives there at the beginning of astronomical summer, about the 22d of June, and the southern one is called the winter solstice, because the sun arrives there at the beginning of the astronomical winter, about the 22d of December. The name solstice comes from two Latin words meaning "the standing still of the sun," because when it is at the solstitial points its apparent course through the sky is for several days nearly horizontal and its declination changes very slowly.

Now, just as there are two opposite points in the sky at equal distances from the equator, which mark the poles of the imaginary axis about which the celestial sphere makes its diurnal revolution, so there are two opposite points at equal distances from the ecliptic which mark the poles of the imaginary axis about which the yearly revolution of the sun takes place. These are called the poles of the ecliptic, and they are situated 23½° from the celestial poles—a distance necessarily corresponding with the inclination of the ecliptic to the equator. The northern pole of the ecliptic is in the constellation Draco, which you may see any night circling round the North Star, together with the Great Dipper and Cassiopeia.

9. Celestial Latitude and Longitude. We have seen that the celestial sphere is marked with imaginary circles resembling the circles of latitude and longitude on the earth, and that in both cases the circles are used for a similar purpose, viz., to determine the location of objects, in one case on the globe of the earth and in the other on the sphere of the heavens. It has also been explained that what corresponds to latitude on the celestial sphere is called declination, and what corresponds to longitude is called right ascension. It happens, however, that these same terms, latitude and longitude, are also employed in astronomy. But, unfortunately, they are based upon a different set of circles from that which has been described, and they do not correspond in the way that right ascension and declination do to terrestrial longitude and latitude. A few words must therefore be devoted to celestial latitude and longitude, as distinguished from declination and right ascension.

Celestial latitude and longitude then, instead of being based upon the equator and the poles, are based upon the ecliptic and the poles of the ecliptic. Celestial latitude means distance north or south of the ecliptic (not of the equator), and celestial longitude means distance from the vernal equinox reckoned along the ecliptic (not along the equator). Celestial longitude runs, the same as right ascension, from west toward east, but it is reckoned in degrees instead of hours. Celestial latitude is measured the same as declination, but along circles running through the poles of the ecliptic instead of the celestial poles, and drawn perpendicular to the ecliptic instead of to the equator. Circles of celestial latitude are drawn parallel to the ecliptic and centring round the poles of the ecliptic, and meridians of celestial longitude are drawn through the poles of the ecliptic and perpendicular to the ecliptic itself. The meridian of celestial longitude that passes through the two equinoxes is the ecliptic prime meridian. This intersects the equinoctial colure at the equinoctial points, making with it an angle of 23½°. The solstitial colure,

which it will be remembered runs round the celestial sphere half-way between
the equinoxes, is perpendicular to the ecliptic as well as to the equator, and so
is common to the two systems of circles. It passes alike through the celestial
poles and the poles of the ecliptic. It will also be observed that the vernal
equinox is common to the two systems of co-ordinates, because it lies at one of
the intersections of the ecliptic and the equator. In passing from one system
to the other, the astronomer employs the methods of spherical trigonometry.

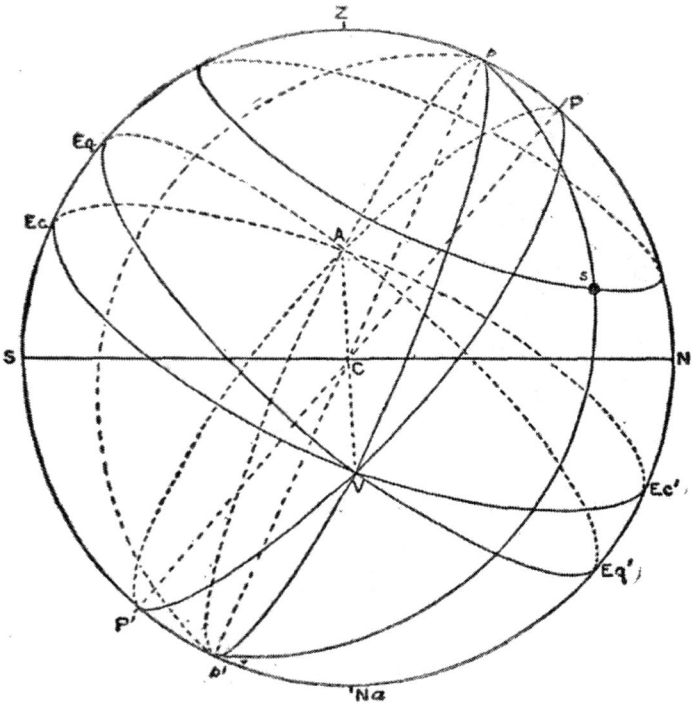

Fig. 4. The Ecliptic and Celestial Latitude and Longitude.

C, as in the other figures, is the place of the observer and Z is the zenith,
but to avoid complication of details the circle of the horizon is not drawn,
only the north-and-south line, N C S, being shown.

Eq Eq is the equator.

Ec Ec is the ecliptic.

P and P are the celestial poles.

p and p are the poles of the ecliptic.

Na is the nadir.

V is the vernal equinox, and A the autumnal equinox.

The circle through s, parallel to the ecliptic, is a latitude circle.

The circle p s p' is the ecliptic meridian of the star s.

The circle P V P A is the equinoctial colure.

The circle p V p A is the prime ecliptic meridian.

The arc of the ecliptic meridian contained between the ecliptic and s measures the star's latitude.

The arc of the ecliptic contained between V and the point where the ecliptic meridian p s p meets the ecliptic (or the angle V p s) measures the star's longitude east from V, the vernal equinox.

10. The Zodiac and the Precession of theEquinoxes. The next thing with which we must make acquaintance is the zodiac. We have learned that the ecliptic is a great circle of the celestial sphere inclined at an angle of 23½° to the equator, and crossing the latter at two opposite points called the equinoxes, and that the sun in its annual journey round the sky follows the circle of the ecliptic. Consequently, the place which the sun occupies at any time must be somewhere on the course of the ecliptic. The fact has been mentioned that as seen from the sun the earth would appear to travel round the ecliptic, whence the ecliptic may be regarded as the projection of the earth's orbit, or path, against the background of the heavens. But, besides the earth there are seven other large planets, Mercury, Venus, Mars, Jupiter, Saturn, Uranus, and Neptune, which, like it, revolve round the sun, some nearer and some farther away. Now, the orbits of all of these planets lie in planes nearly coincident with that of the earth's orbit. None of them is inclined more than 7° from the ecliptic and most of them are inclined only one or two degrees. Consequently, as we watch these planets moving slowly round in their orbits we find that they are always quite close to the circle of the ecliptic. This fact shows that the solar system, *i. e.*, the sun and its attendant planets, occupies a disk-shaped area in space, the outlines of which would be like those of a very thin round cheese, with the sun in the centre. The ecliptic indicates the median plane of this imaginary disk. The moon, too, travels nearly in this common plane, its orbit round the earth being inclined only a little more than 5° to the ecliptic.

Even the early astronomers noticed these facts, and in ancient times they gave to the apparent road round the sky in which the sun and planets travel, in tracks relatively as close together as the parallel marks of wheels on a highway, the name zodiac. They assigned to it a certain arbitrary width, sufficient to include the orbits of all the planets known to them. This width is 8° on each side of the circle of the ecliptic, or 16° in all. They also divided the ring of the zodiac into twelve equal parts, corresponding with the number of months in a year, and each part was called a sign of the zodiac. Since there are 360° in a circle, each sign of the zodiac has a length of just 30°. To indicate the course of the zodiac to the eye, its inventors observed the constellations lying along it, assigning one constellation to each sign. Beginning at the vernal equinox, and running round eastward, they gave to these zodiacal constellations, as well as to the corresponding signs, names drawn from fancy resemblances of the figures formed by the stars to men, animals, or other objects. The first sign and constellation were called Aries, the Ram, indicated by the symbol ;

the second, Taurus, the Bull, ; the third, Gemini, the Twins, ; the fourth, Cancer, the Crab, ; the fifth, Leo, the Lion, ; the sixth, Virgo, the Virgin, ; the seventh, Libra, the Balance, ; the eighth, Scorpio, the Scorpion, ; the ninth, Sagittarius, the Archer, ; the tenth, Capricornus, the Goat, ; the eleventh, Aquarius, the Water-Bearer, ; and the twelfth, Pisces, the Fishes, . The name zodiac comes from a Greek word for animal, since most of the imaginary figures formed by the stars of the zodiacal constellations are those of animals. The signs and their corresponding constellations being supposed fixed in the sky, the planets, together with the sun and the moon, were observed to run through them in succession from west to east.

When this system was invented, the signs and their constellations coincided in position, but in the course of time it was found that they were drifting apart, the signs, whose starting point remained the vernal equinox, backing westward through the sky until they became disjoined from their proper constellations. At present the sign Aries is found in the constellation next west of its original position, viz., Pisces, and so on round the entire circle. This motion, as already intimated, carries the equinoxes along with the signs, so that the vernal equinox, which was once at the beginning of the constellation Aries (as it still is at the beginning, or "first point," of the *sign* Aries), is now found in the constellation Pisces.

To explain the shifting of the signs of the zodiac on the face of the sky we must consider the phenomenon known as the precession of the equinoxes, which is one of the most interesting things in astronomy. Let us refer again to the fact that the axis of the earth's daily rotation is inclined 23½° from a perpendicular to the plane of its yearly revolution round the sun, from which it results that the ecliptic is tipped at the same angle to the plane of the equator. Thus the sun, moving in the ecliptic, appears half the year above (or north of) the equator, and half the year below (or south of) it, the crossing points being the two equinoxes. Now, this inclination of the earth's axis is the key to the explanation we are seeking. The direction in which the axis lies in space is a *fixed* direction, which can be changed only by some outside force interfering. What we mean by this will become clearer if we think of the earth's axis as resembling the peg of a top, or the axis of a gyroscope. When a top is spinning smoothly, with its peg vertical, the peg will remain vertical as long as the spin is not diminished, and no outside force interferes. So, too, the axis of the spinning-wheel of a gyroscope keeps pointing in the same direction so persistently that the wheel is kept from falling. If it is so mounted that it is free to move in any direction, and if then you take the instrument in your hand and turn round with it, the axis will adjust itself in such a manner as to retain its original direction in space. This tendency of a rotating body to keep its axis of rotation fixed applies equally to the earth, whose axis, also, maintains a constant direction in space, except for a slow change produced by outside forces, which change constitutes the phenomenon of the precession of the equinoxes.

We cannot too often recall the fact that the axis of the earth is coincident in direction with that of the celestial sphere, so that the earth's poles are situated directly under the celestial poles. But the poles of the ecliptic are 23½° aside from the celestial poles. If the direction of the earth's axis and with it that of the celestial sphere, did not change at all, then the celestial poles and the poles of the ecliptic would always retain the same relative positions in the sky; but, in fact, an exterior force, acting upon the earth, causes a gradual change in the direction of its axis, and in consequence of this change the celestial poles, whose position depends upon that of the earth's poles, have a slow motion of revolution about the poles of the ecliptic, in a circle of 23½° radius. The force which produces this effect is the attraction of the sun and the moon upon the protuberant part of the earth round its equator. If the earth were a perfect sphere, this force could not act, or would not exist, but since the earth is an oblate spheroid, slightly flattened at the poles, and bulged round the equator, the attraction acts upon the equatorial protuberance in such a way as to strive to pull the earth's axis into an upright position with respect to the plane of the ecliptic. But, in consequence of its spinning motion, the earth resists this pull, and tries, so to speak, to keep the inclination of its axis unchanged. The result is that the axis swings slowly round while maintaining nearly the same inclination to the plane of the ecliptic.

Here, again, we may employ the illustration of a top. If the peg of the top is tipped a little aside, so that the attraction of gravitation would cause the top to fall flat on the table if it were not spinning, it will, as long as it continues to spin, swing round and round in a circle instead of falling. We cannot enter into a mathematical explanation of this phenomenon here, but the reader will find a clear popular account of the whole matter in Prof. John Perry's little book on *Spinning Tops.* It is sufficient here to say that the attraction of gravitation, tending to make the top fall, but really causing the peg to turn round and round, resembles, in its effect, the attraction of the sun and the moon upon the equatorial protuberance of the earth, which makes the earth's axis turn round in space.

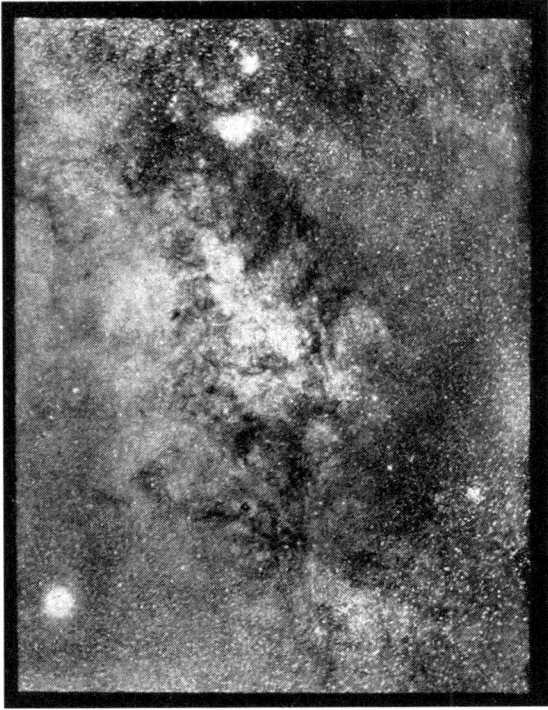

The Milky Way about Chi Cygni

Photographed at the Lick Observatory by E. E. Barnard, with the six-inch Willard lens.

Observe the cloud-like forms.

Now, as we have said, this slow swinging round of the axis of the earth produces the so-called precession of the equinoxes. In a period of about 25,800 years, the axis makes one complete swing round, so that in that space of time the celestial poles describe a revolution about the poles of the ecliptic, which remain fixed. But since the equator is a circle situated half-way between the poles, it is evident that it must turn also. To illustrate this, take a round flat disk of tin, or pasteboard, to represent the equator and its plane, and perpendicularly through its centre run a straight rod to represent the axis. Put one end of the axis on the table, and, holding it at a fixed inclination, turn the upper end round in a circle. You will see that as the axis thus revolves, the disk revolves with it, and if you imagine a plane, parallel to the surface of the table, passing through the centre of the disk at the point where the rod pierces it, you will perceive that the two opposite points, where the edge of the disk intersects this imaginary plane, revolve with the disk. In one position of the axis, for instance, these points may lie in the direction of the north-and-south sides of the room. When you have revolved the axis, and with it the disk, one quarter way round, the points will lie toward the east and west sides of the room. When you have produced a half revolution they will once more lie toward the north-and-south, but now the direction of the slope of the disk will be the reverse of that which it had at the beginning. Finally, when the revolution is completed, the two points will again lie north-and-south and the slope of the disk will be in the same direction as at the start. In this illustration the disk stands for the plane of the celestial equator, the rod for the axis of the celestial sphere, the imaginary plane parallel to the surface of the table for the plane of the ecliptic, and the two opposite points where this plane is intersected by the edge of the disk for the equinoxes. The motion of these points as the inclined disk revolves represents the precession of the equinoxes. This term means that the direction of the motion of the equinoxes, as they shift their place on the ecliptic, is such that they seem to precess, or move forward, as if to meet the sun in its annual journey round the ecliptic. The direction is from east to west, and thus the zodiacal signs are carried farther and farther westward from the constellations originally associated with them; for these signs, as we have said, are so arranged that they begin at the vernal equinox, and if the equinox moves, the whole system of signs must move with it. The amount of the motion is about 50 .2 per year, and since there are 1,296,000 in a circle, simple division shows that the time required for one complete revolution of the equinoxes must be, as already stated in reference to the poles, about 25,800 years. A little over 2000 years ago the signs and the constellations were in accord; it follows, then, that about 23,800 years in the future, they will be in accord again. In the meanwhile the signs will have backed entirely round the circle of the ecliptic.

The attentive reader will perceive that the precession of the equinoxes, with its attendant revolution of the celestial poles round the poles of the ecliptic, must affect the position of the North Star. We have already said that that star only *happens* to occupy its present commanding position in the sky. The star itself is motionless, or practically so, with regard to the earth, and it is the north pole that changes its place. At the present time the pole is about 1° 10, from the North Star, in the direction of the Great Dipper, and it is slowly drawing nearer so that in about 200 years it will be less than half a degree from the star. After that the precessional motion will carry the pole in a circle departing farther and farther from the star, until the latter will have entirely lost its importance as a guide to the position of the pole. It happens, however, that several other conspicuous stars lie near this circle. One of these is Thuban, or Alpha Draconis (not now as bright as it once was), and this star at the time when it served as an indicator of the place of the pole, some 4600 years ago, was connected with a very romantic chapter in the history of astronomy. In the great pyramid of Cheops in Egypt, there is a long passage leading straight toward the north from a chamber cut deep in the rock under the centre of the pyramid, and the upward slope of this passage is such that it is believed to have been employed by the Egyptian astronomer-priests as a kind of telescope-tube for viewing the then pole star, and observing the times of its passage over the meridian—for even the North Star, since it is not *exactly* at the pole, revolves every twenty-four hours in a tiny circle about it, and consequently crosses the meridian twice a day, once above and once beneath the true pole.

About 11,500 years in the future, the extremely brilliant star Vega, or Alpha Lyræ, will serve as a pole star, although it will not be as near the pole as the North Star now is. At that time the North Star will be nearly 50° from the pole. In about 21,000 years the pole will have come round again to the neighbourhood of Alpha Draconis, the star of the pyramid, and in about 25,800 years the North Star will have been restored to its present prestige as the apparent hub of the heavens.

One curious irregularity in the motion of the earth's poles must be mentioned in connection with the precession of the equinoxes. This is a kind of "nodding," known as nutation. It arises from variation in the effect of the attraction of the sun and the moon, due to the varying directions in which the attraction is exercised. As far as the sun is concerned, the precession is slower near the time of the equinoxes than in other parts of the year; in other words, it is most rapid in mid-summer and mid-winter when one or the other of the poles is turned sunward. A similar, but much larger, change takes place in the effect of the moon's attraction owing to the inclination of her orbit to the ecliptic. During about nine and a half years, or half the period of revolution of her nodes (see Part III, Section 4), the moon tends to hasten the precession, and during the next nine and a half years to retard it. The general effect of the combination of these irregularities is to cause the earth's poles to describe a slightly waving curve instead of a smooth circle round the poles of the ecliptic. There are about 1400 of these "waves," or "nods," in the motion of the poles in the course of their 26,000-year circuit. In accurate observation the astronomer is compelled to take account of the effects of nutation upon the apparent places of the stars.

A very remarkable general consequence of the change in the direction of the earth's axis will be mentioned when we come to deal with the seasons.

The Great Southern Star-Cluster Centauri

Photographed by S. I. Bailey at the South American Station of Harvard Observatory.

Note the streaming of small stars around the cluster. The cluster itself is globular and its stars are too numerous to be counted, or even to be separately distinguished in the central part.

PART II. THE EARTH.

PART II.   THE EARTH.

1. Nature, Shape, and Size of the Earth. The situation of the earth in the universe has been briefly described in Part I; it remains now to see what the earth is in itself, and what are some of the principal phenomena connected with it as a celestial body inhabited by observant and reasoning beings.

We know by ordinary experience that the earth is composed of rock, sand, soil, etc., and generally covered, where there is no running or standing water in the form of rivers, lakes, or seas, with vegetation, such as trees and grass. Further experience teaches us that the earth is very large, and that its surface is divided into wide areas of land and of water. The largest bodies of water, the oceans, taken all together, cover about 72 per cent., or nearly three-quarters of the entire surface of the earth. Investigations carried as far down as we can go show that the interior of the earth consists of various kinds of rock, in which are contained many different kinds of metals. While there is reason for thinking that a high degree of temperature prevails deep in the earth, yet it appears evident, for other reasons, that, taken as a whole, it is solid and very rigid throughout. By methods, the history and description of which we have not here sufficient space to give, it has been proved that the earth is, in form, a globe, or more strictly an ellipsoid, slightly drawn in at the poles and swollen round the equator. The polar diameter is 7899 miles, and the equatorial diameter 7926 miles, the difference amounting to only 27 miles. Thus, for ordinary purposes, we may regard the earth as being a true sphere. The level of its surface, however, is varied by hills and mountains, which, though insignificant in comparison with the size of the whole earth, are enormous when compared with the structures of human hands. The loftiest known mountain on the earth, Mt. Everest in the Himalayas, has an elevation of 29,000 feet above sea-level, and the deepest known depression of the ocean bottom, near the island of Guam in the Pacific, sinks 31,614 feet below sea-level. Thus, the apex of the highest mountain is about eleven and a half miles in vertical elevation above the bottom of the deepest pit of the sea—a distance very considerably less than half the difference between the equatorial and polar diameters of the earth.

It is believed that at the beginning of its history the earth was a molten mass, or perhaps a mass of hot gases and vapours like the sun, and that it assumed its present shape in obedience to mechanical laws, as it cooled off. The rotation caused it to swell round the equator and draw in at the poles.

The outer part of the earth is called its crust, and geology shows that this has been subject to violent changes, such as upheavals and subsidences, and that in many places sea and land have interchanged places, probably more than once. Geology also shows that the rocks of the earth's crust are filled with the remains, or fossils, of plants and animals differing from those now existing, though related to them, and that many of these must have lived millions of years ago. Thus we see that the earth bears marks of an immense antiquity, and that it was probably inhabited during vast ages before the race of man had been developed. The origin of life upon the earth is unknown.

2. The Attraction of Gravitation. Among the phenomena of life upon the earth, which are so familiar that only thoughtful persons see anything to wonder at in them, is what we call the "weight" of bodies. Every person feels that he is held down to the ground by his weight, and he knows that if he drops a heavy body it will fall straight toward the ground. But what is this weight which causes everything either to rest upon the earth or to fall back to it if lifted up and dropped? The answer to this question involves a principle, or "law," which affects the whole universe, and makes it what we see it. This principle is one of the great foundation stones of astronomy. It is called the law of gravitation, the word gravitation being derived from the Latin *gravis*, "heavy." Briefly stated, the law is that every body, or every particle of matter, attracts, or strives to draw to itself, every other body, or particle of matter. This force is called the attraction of gravitation. A large body possesses more attractive force than a small one, in proportion to the mass, or quantity of matter, that it contains. The earth, being extremely large, holds all bodies on its surface with a force proportionate to its great mass. This explains why we possess what we call weight, which is simply the effect of the attraction of the earth upon our bodies. A large body is heavier, or drawn with more force by the earth, than a small one (composed of the same kind of matter), because it has a greater mass. The body really attracts the earth as much as the earth attracts the body, but the amount of motion caused by the attraction is proportional to the respective masses of the attracting bodies, and since the mass of the earth is almost infinitely great in comparison with that of any body that we can handle, the motion which the latter imparts to the earth is imperceptible, and it is the small body only that is seen to move under the force of the attraction.

Now we are going to see how vastly important in its effects is the fact that the earth is spherical in form. Sir Isaac Newton, who first worked out mathematically the law of gravitation, proved that a spherical body attracts, and is attracted, as if its entire mass were concentrated in a point at its centre. From this it follows that the attraction of the earth is exercised just as if the whole attractive force emanated from a middle point, and, that being so, the effect of the attraction is to draw bodies from all sides toward the centre of the earth. This explains why people on the opposite side of the earth, or under our feet, as we say, experience the same attractive force, or have the same weight, that we do. All round the earth, no matter where they may be situated, objects are drawn toward the centre. If at any point on the earth you suspend a plumb-line, and then, going one quarter way round, suspend another plumb-line, each of the lines will be vertical at the place where it hangs, and yet, the directions of the two lines will be at right angles to one another, since both point toward the centre of the earth.

Knowing the manner in which the earth attracts, we have the means of determining its entire mass, or, as it is sometimes called, the weight of the earth. The principle on which this is done is easily understood, Suppose, for instance, that a small ball of lead, of known weight, is brought near a large ball, and delicately suspended in such a way that, by microscopic observation, the movement imparted by the attraction of the large ball can be measured. The force required to produce this movement can be compared with the force of the earth's attraction which produces the weight of the ball, and thus the ratio of the mass of the earth to that of the ball is determined. The total mass of the earth has been found to be equivalent to a "weight" of about 6,500,000,000,000,000,000,000 tons. The mean density of the earth compared with that of water is found to be about 5½, that is to say, the earth weighs 5½ times as much as a globe of water of equal size.

Newton did not stop with showing the manner of the earth's attraction upon bodies on or near its surface; he proved that the earth attracted the moon also, and thus retained it in its orbit. To understand this we must notice another fact concerning the manner in which gravitation acts. Its force varies with distance. Experiment followed by mathematical demonstration, has proved that the variation of the attraction is inversely proportional to the square of the distance. This simply means that if the distance between the two bodies concerned is doubled, the force of attraction will be diminished four times, 4 being the square of 2; and that if the distance is halved, the force will be increased fourfold. Increase the distance three times, and the force diminishes nine times; diminish the distance three times, and the force increases nine times, because 9 is the square of 3, and, as we have said, the force varies inversely, or contrarily, to the change of distance. Knowing this, Newton computed what the force of the earth's attraction must be on the moon, and he found that it was just sufficient to keep the latter moving round and round the earth. But why does not the moon fall directly to the earth? Because the moon had originally another motion across the direction of the earth's attraction. How it got this motion is a question into which we cannot here enter, but, if it were not attracted by the earth (or by the sun), the moon would travel in a straight line through space, like a stone escaping from a sling. The force of the attraction is just sufficient to make the moon move in an orbital path about the earth.

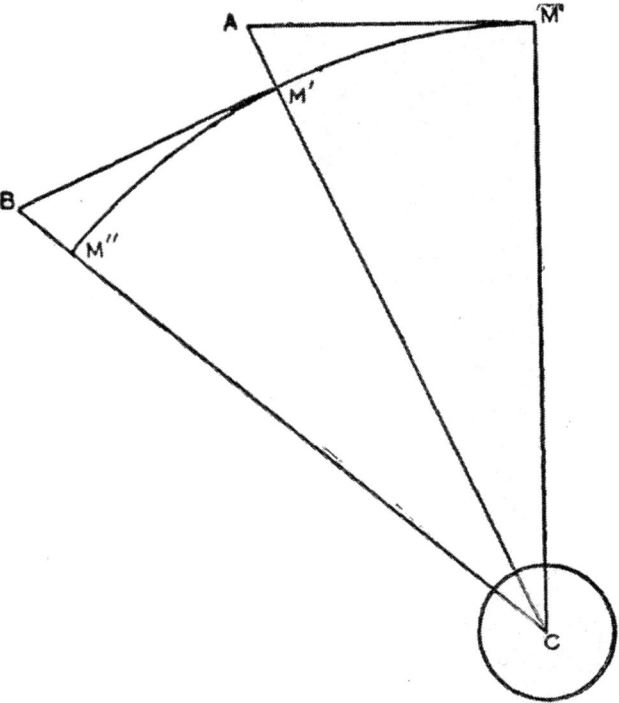

Fig. 5. How the Earth Controls the Moon.

Let C be the centre of the earth and M that of the moon. Suppose the moon to be moving in a straight line at such a velocity that it will, if not interfered with, go to A in one day. Now suppose the attraction of the earth to act upon it. That attraction will draw it to M . Again suppose that at M the moon were suddenly released from the earth's attraction; it would then shoot straight ahead to B in the course of the next day. But, in fact, the earth's attraction acts continually, and in the second day the moon is drawn to M . In other words the moon is all the time falling away from the straight line that it would pursue but for the earth's attraction, and yet it does not get nearer the earth but simply travels in an endless curve round it.

The same principle was extended by Newton to explain the motion of the earth around the sun. The force of the sun's attraction, calculated in the same way, can be shown to be just sufficient to retain the earth in its orbit and prevent it from travelling away into space. And so with all the other planets which revolve round the sun. And this applies throughout the universe. There are certain so-called double, or binary, stars, which are so close together that their attraction upon one another causes them to revolve in orbits about their common centre. In truth, all the stars attract the earth and the sun, but the force of this attraction is so slight on account of their immense distance that we cannot observe its effects. The reader who wishes to pursue this subject of gravitational attraction should consult more extensive works, such as Prof. Young's *General Astronomy*, or Sir George Airy's *Gravitation*.

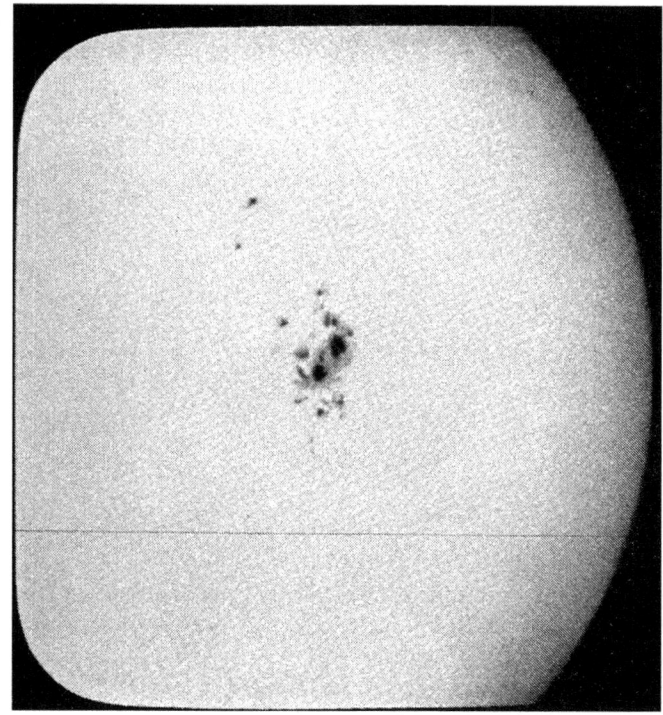

Photograph of a Group of Sun-spots
Similar groups are frequently seen during periods of sun-spot maximum.

3. The Tides. The tides in the ocean are a direct result of the attraction of gravitation. They also involve in an interesting way the principle that a spherical body, like the earth, attracts and is attracted as if its entire mass were concentrated at its centre. The cause of the tides is the difference in the attraction of the sun and moon upon the body of the earth as a rigid sphere, and upon the water of the oceans, as a fluid envelope whose particles, while not free to escape from the earth, are free to move, or slide, among one another in obedience to varying forces. The difference of the force of attraction arises from the difference of distance. Since the moon, because of her relative nearness, is the chief agent in producing tides we shall, at first, consider her tidal influence alone. The diameter of the earth is, in round numbers, 8000 miles; therefore, its radius is 4000 miles. From this it follows that the centre of the earth is 4000 miles farther from the moon than that side of the earth which is toward her at any time, and 4000 miles nearer than the side which is away from her. Consequently, her attraction must be stronger upon the water of the ocean lying just under her than upon the centre of the earth, and it must also be stronger upon the centre of the earth than upon the water of the ocean lying upon the side which is farthest from her. The result of these differences in the force of the moon's attraction is that the water directly under her tends away from the centre of the earth, while, on the other hand, the earth, considered as a solid sphere, tends away from the water on the side opposite to that where the moon is, and these combined tendencies cause the water to rise, with regard to its general level, in two protuberances, situated on opposite sides of the earth. These we call tides.

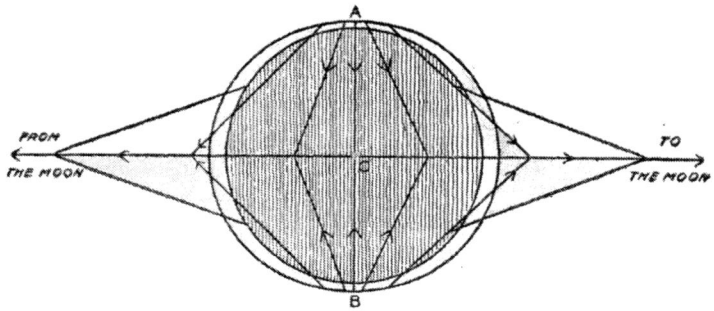

Fig. 6. The Tidal Force of the Moon.

The solid earth is represented surrounded by a shell of water. The water on the side toward the moon is more attracted than the centre of the earth, C; the water on the opposite side is less attracted. The lines of force from the moon to the parts of the water lying toward A and B are inclined to the direct line between the centres of the earth and the moon, and the forces acting along these lines tend to draw the water in the directions shown by the arrow points. These are resultants of the horizontal and vertical components of the moon's attraction at the corresponding points on the earth, and the force acting along them tends to increase the weight of the water wherever the lines are inclined more toward the centre of the earth than toward the moon. On the side opposite the moon the same effects are produced in reverse, because on that side the general tendency is to draw the earth away from the water. Consequently if the earth did not rotate, and if it were surrounded with a complete shell of water, the latter would be drawn into an ellipsoidal shape, with the highest points under and opposite to the moon, and the lowest at the extremities of the diameter lying at right angles to the direction of the moon.

Some persons, when this statement is made, inquire: "Why, then, does not the moon take the water entirely away from the earth?" The answer is, that the effect of the tidal force is simply to diminish very slightly the weight of the water, or its tendency towards the earth's centre, but not to destroy, or overmaster, the gravitational control of the earth. The water retains nearly all its weight, for the tidal force of the moon diminishes it less than one part in 8,000,000. Still, this slight diminution is sufficient to cause the water to swell a little above its general level, at the points where it feels the effect of the tidal force. On the other hand, around that part of the earth which is situated half-way between the two tides, or along a diameter at right angles to the direction of the moon, the latter's attraction *increases* the weight of the water, *i. e.*, its tendency toward the earth's centre (see Fig. 6). Perhaps this can better be understood, if we imagine the earth to be entirely liquid. In that case the difference in the force of the moon's attraction with difference of distance would be manifested in varying degrees throughout the earth's whole frame, and the result would be to draw the watery globe out into an ellipsoidal figure, having its greatest diameter in the line of the moon's attraction, and its smallest diameter at right angles to that line. The proportions of the ellipsoid would be such that the forces would be in equilibrium.

Owing to a variety of causes, such as the rotation of the earth on its axis, which carries the water rapidly round with it; the inertia of the water, preventing it from instantly responding to the tidal force; the irregular shape of the oceans, interrupted on all sides by great areas of land; their varying depth, producing differences of friction, and so on, the tidal waves do not appear directly under, or directly opposite to, the moon, and the calculation of the course and height of the actual tides, at particular points on the earth, becomes one of the most difficult problems in astronomical physics.

We now turn to consider the effects of the sun's tidal force in connection with that of the moon. This introduces further complications. The solar tides are only about two-fifths as high as the lunar tides, but they suffice to produce notable effects when they are either combined with, or act in opposition to, the others. They are combined twice a month—once when the moon is between the earth and the sun, at the time of new moon; and again when the moon is in opposition to the sun, at the time of full moon. In these two positions the attractions of the sun and the moon must, so to speak, act together, with the result that the tides produced by them blend into a single greater wave. This combination produces what are called spring tides, the highest of the month. When, on the other hand, the moon is in a position at right angles to the direction of the sun, which happens at the lunar phases named first and last quarters, the solar and the lunar tides have their crests 90° apart, and, in a sense, act against one another, and then we have the neap tides, which are the lowest of the month.

Without entering into a demonstration, it may here be stated as a fact to be memorised, that the tidal force exerted by any celestial body varies inversely as the *cube* of the distance. This is the reason why the sun, although it exceeds the moon in mass more than 25,000,000 times, and is situated only about 400 times as far away from the earth, exercises comparatively so slight a tidal force on the water of the ocean. If the tidal force varied as the *square* of the distance, like the ordinary effects of gravitation, the tides produced by the sun would be more than 150 times as high as those produced by the moon, and would sweep New York, London, and all the seaports of the world to destruction. In that case it might be possible, by delicate observations, to detect a tidal effect produced upon the oceans of the earth by the planet Jupiter.

4. The Atmosphere. The solid globe of the earth is enveloped in a mixture of gases, principally oxygen and nitrogen, which we call the air, or the atmosphere, and upon whose presence our life and most other forms of life depend. The atmosphere is retained by the attraction of the earth, and it rotates together with the earth. If this were not so—if the atmosphere stood fast while the earth continued to spin within it—a terrific wind would constantly blow from the east, having a velocity at the equator of more than a thousand miles an hour.

Exactly how high the atmosphere extends we do not know—it may not have any definite limits—but we do know that its density rapidly diminishes with increase of height above the ground, so that above an elevation of a few miles it becomes so rare that it would not support human life. The phenomena of meteors, set afire by the friction of their swift rush through the upper air, prove, however, that there is a perceptible atmosphere at an elevation of more than a hundred miles.

From an astronomical point of view, the most important effect of the presence of the atmosphere is its power of refracting light. By refraction is meant the property possessed by every transparent medium of bending, under particular circumstances, the rays of light which enter it out of their original course. The science of physics teaches us that if a ray of light passes from any transparent medium into another which is denser, and if the path of this ray is not perpendicular to the surface of the second medium, it will be turned from its original course in such a way as to make it more nearly perpendicular. Thus, if a ray of light passes from air into water at a certain slope to the surface, it will, upon entering the water, be so changed in direction that the slope will become steeper. Only if it falls perpendicularly upon the water will it continue on without change of direction. Conversely a ray passing from a denser into a rarer medium is bent away from a perpendicular to the surface of the first medium, or its slope becomes less. This explains why, if we put a coin in a bowl, with the eye in such a position that it cannot see the coin over the edge, and then fill the bowl with water, the coin seems to be lifted up into sight. Moreover, if any transparent medium increases in density with depth, the amount of refraction will increase as the ray goes deeper, and the direction of the ray will be changed from a straight line into a curve, tending to become more and more perpendicular.

Now all this applies to the atmosphere. If a star is seen in the zenith, its light falls perpendicularly into the atmosphere and its course is not deviated, or in other words there is no refraction. But if the star is somewhere between the zenith and the horizon, its light falls slopingly into the atmosphere, and is subject to refraction, the amount of bending increasing with approach to the horizon. Observation shows that the refraction of the atmosphere, which is zero at the zenith, increases to about half a degree (and sometimes much more, depending upon the state of the air), near the horizon. It follows that a celestial object seen near the horizon will ordinarily appear about half a degree above its true place. Since the apparent diameters of the sun and the moon are about half a degree, when they are rising or setting they can be seen on the horizon before they have really risen above it, or after they have really sunk below it. Tables of refraction at various altitudes have been prepared, and they have to be consulted in all exact observations of the celestial bodies.

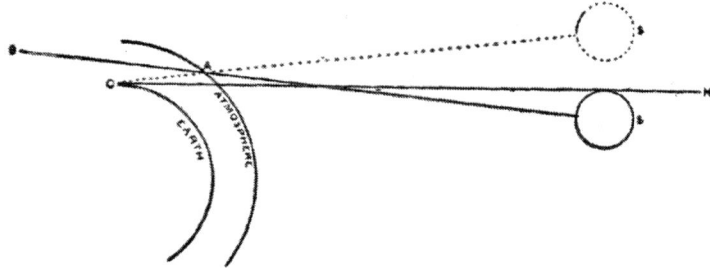

Fig. 7. Refraction.

Suppose an observer situated at O on the earth. The sun, at S, has sunk below the level of his horizon, O H, but since the sun sends out rays in all directions there will be some, such as S A B, which will strike the atmosphere at A, and the refraction, tending to make the ray more nearly perpendicular to the surface of the atmosphere, will, instead of allowing it to go on straight over the observer's head to B, bend it down along the dotted line A O, and the observer will see the sun as if it lay in the direction of the dotted line O A S , which places the sun apparently above the horizon.

5. Dip of the Horizon. Another correction which has to be applied in many observations depends upon the sphericity of the earth. We have described the rational horizon, and pointed out how it differs from the sensible horizon. We have also said that at sea the sensible horizon nearly accords with the rational horizon (see Part I, Sect. 3). But the accord is not complete, owing to what is called the dip of the horizon. In fact, the sea horizon lies below the rational horizon by an amount varying with the elevation of the eye above the surface. Geometry enables us to determine just what the dip of the horizon must be for any given elevation of the eye. A rough and ready rule, which may serve for many purposes, is that the square root of the elevation of the eye in feet equals the dip of the horizon in minutes of arc, or of angular measure. The reader will readily see that the dip of the horizon is a necessary consequence of the rotundity of the earth. It is because of this that, as a ship recedes at sea her hull first disappears below the horizon, and then her lower sails, and finally her top-sails. The use of a telescope does not help the matter, because a telescope only *sees straight*, and cannot bend the line of sight over the rim of the horizon. Atmospheric refraction, however, enables us to see an object which would be hidden by the horizon if there were no air. In navigation, which, as a science, is an outgrowth of astronomy, these things have to be carefully taken into account.

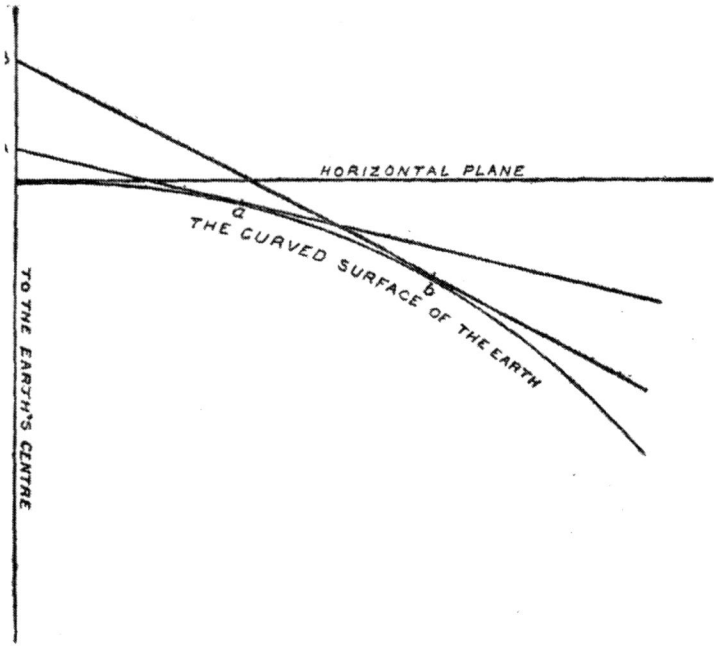

Fig. 8. Dip of the Horizon.

It is to be remembered that it is the *sensible* horizon which dips, and not the *rational* horizon. The sensible horizon of the observer at the elevation A dips below the horizontal plane and he sees round the curved surface as far as a; in other words his skyline is at a. The observer at the elevation B has a sensible horizon still more inclined and he sees as far as b. If the observation were made from an immense height the observer would see practically half round the earth

just as we see half round the globe of the moon.

Polar Streamers of the Sun, Eclipse of 1889

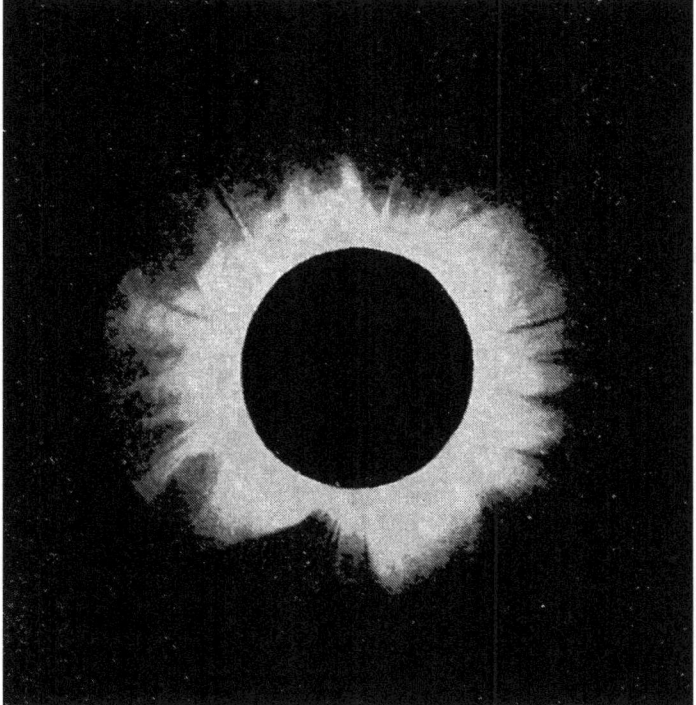

The Solar Corona at the Eclipse of 1871 From drawings.

6. Aberration. A few words must be said about the phenomenon known as aberration of light. This is an apparent displacement of a celestial object due to the motion of the earth in its orbit. It is customary to illustrate it by imagining oneself to be in a shower of rain, whose drops are falling vertically.

In such a case, if a person stands fast the rain will descend perpendicularly upon his head, but if he advances rapidly in any direction he will feel the drops striking him in the face, because his own forward motion is compounded with the downward motion of the rain so that the latter seems to be descending slantingly toward him. The same thing happens with the light falling from the stars. As the earth advances in its orbit it seems to meet the light rays, and they appear to come from a direction ahead of the flying earth. The result is that, since we see a star in the direction from which its light seems to come, the star appears in *advance* of its real position, or of the position in which we would see it if the earth stood fast. The amount by which the position of a star is shifted by aberration depends upon the ratio of the earth's velocity to the velocity of light. In round numbers this ratio is as 1 to 10,000. The motion of the earth being in a slightly eccentric ellipse, the stars describe corresponding, but very tiny, ellipses once every year upon the background of the sky. But the precise shape of the ellipse depends upon the position of the star on the celestial sphere. If it is near one of the poles of the ecliptic, it will describe an annual ellipse which will be almost a circle, its greater diameter being 41 of arc. If it is near the plane of the ecliptic, it will describe a very eccentric ellipse, but the greater diameter will always be 41 , although the shorter diameter may be immeasurably small. The effects of aberration have to be allowed for in all careful astronomical observation either of the sun or the stars. This is done by reducing the apparent place of the object to the place it would have if it were seen at the centre of its annual ellipse.

7. Time. Without astronomical observations we could have no accurate knowledge of time. The basis of the measurement of time is furnished by the rotation of the earth on its axis. We divide the period which the earth occupies in making one complete turn into twenty-four equal parts, or hours. The ascertainment of this period, called a day, depends upon observations of the stars. Suppose we see a certain star exactly on the meridian at some moment; just twenty-four hours later that star will have gone entirely round the sky, and will again appear on the meridian. The revolving heavens constitute the great clock of clocks, by whose movements all other clocks are regulated. We know that it is not the heavens which revolve, but the earth which rotates, but for convenience we accept the appearance as a substitute for the fact. The rotation of the earth is so regular that no measurable variation has been found in two thousand years. We have reasons for thinking that there must be a very slow and gradual retardation, owing principally to the braking action of the tides, but it is so slight that we cannot detect it with any means at present within our command.

In Part I it was shown how the passage across the meridian of the point in the sky called the vernal equinox serves to indicate the beginning of the astronomical "day," but the position of the vernal equinox itself has to be determined by observations on the stars. By means of a telescope, so mounted that it can only move up or down, round a horizontal axis, and with the axis pointing exactly east and west so that the up and down movements of the telescope tube follow the line of the meridian, the moment of passage across the meridian of a star at any altitude can be observed. Observations of this nature are continually made at all great government observatories, such as the observatory at Washington or that at Greenwich, and at many others, and by their means clocks and chronometers are corrected, and a standard of time is furnished to the whole world.

There are, however, three different ways of reckoning time, or, as it is usually said, three kinds of time. One is sidereal time, which is indicated by the passage of stars across the meridian, and which measures the true period of the earth's rotation; another is apparent solar time, which is indicated by the passage of the sun across the meridian; and a third is mean solar time, which is indicated by a carefully regulated clock, whose errors are corrected by star observations. This last kind of time is that which is universally used in ordinary life (the use of sidereal time being confined to astronomy), so it is necessary to explain what it is and how it differs from apparent solar time.

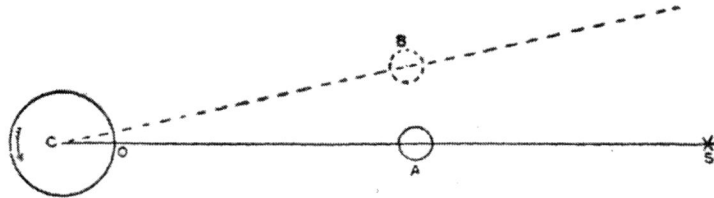

Fig. 9. Sidereal and Solar Time.

C is the centre of the earth, and O the place of an observer on the earth's surface.

Suppose the sun at A to be in conjunction with the star S. Then, at the end of twenty-four sidereal hours, when the earth has made one turn on its axis and the place O has again come into conjunction with the star, the sun, in consequence of its yearly motion in the ecliptic, will have advanced to B, and the earth will have to turn through the angle A C B before O will overtake the sun and complete a solar day; wherefore the solar day is longer than the sidereal.

In the first place, the reason why sidereal time is not universally and exclusively used is because, although it measures the true period of the earth's rotation by the apparent motion of the stars, it does not exactly accord with the apparent motion of the sun; and, naturally, the sun, since it is the source of light for the earth, and the cause of the difference between day and night, is taken for all ordinary purposes, as the standard indicator of the progress of the hours. The fact that it is mid-day, or noon, at any place when the sun crosses the meridian of that place, is a fact of common knowledge, which cannot be ignored. On the other hand, the vernal equinox, which is the "noon mark" for sidereal time, is independent of the alternation of day and night, and may be on the meridian as well at midnight as at mid-day. Before clocks and watches were perfected, the moment of the sun's passage over the meridian was determined by means of a gnomon, which shows the instant of noon by the length of a shadow cast by an upright rod. Since the apparent course of the sun through the sky is a curve, rising from the eastern horizon, attaining its greatest elevation where it meets the meridian, and thence declining to the western horizon, it is evident that the length of the shadow must be least when the sun is on the meridian, or at its maximum altitude. The gnomon, or the sun-dial, gives us apparent solar time. But this differs from sidereal time because, as we saw in Part I, the sun, in consequence of the earth's motion round it, moves about one degree eastward every twenty-four hours, and, since one degree is equal to four minutes of time, the sun rises about four minutes later, with reference to the stars, every morning. Consequently it comes four minutes later to the meridian day after day. Or, to put it in another way, suppose that the sun and a certain star are upon the meridian at the same instant. The star is fixed in its place in the sky, but the sun is not fixed; on the contrary it moves about one degree eastward (the same direction as that of the earth's rotation) in twenty-four hours. Then, when the rotation of the earth has brought the star back to the meridian at the end of twenty-four sidereal hours, the sun, in consequence of its motion, will still be one degree east of the meridian, and the earth must turn through the space of another degree, which will take four minutes, before it can have the sun again upon the meridian. The true distance moved by the sun in twenty-four hours is a little less than one degree, and the exact time required for the meridian to overtake it is 3 min. 56.555 sec. Thus, the sidereal day (period of 24 hours) is nearly four minutes shorter than the solar day.

It would seem, then, that by taking the sun for a guide, and dividing the period between two of its successive passages over the meridian into twenty-four hours, we should have a perfect measure of time, without regard to the stars; in other words, that apparent solar time would be entirely satisfactory for ordinary use. But, unfortunately, the apparent eastward motion of the sun is not regular. It is sometimes greater than the average and sometimes less. This variation is due almost entirely: first, to the fact that its orbit not being a perfect circle the earth moves faster when it is near perihelion, and slower when it is near aphelion; and, second, to the effects of the inclination of the ecliptic to the equator. In consequence, another measure of solar time is used, called *mean* solar time, in which, by imagining a fictitious sun, moving with perfect regularity through the ecliptic, the discrepancies are avoided. All ordinary clocks are set to follow this fictitious, or mean, sun. The result is that clock time does not agree exactly with sun-dial time, or, what is the same thing, apparent solar time. The clock is ahead of the real sun at some times of the year, and behind it at other times. This difference is called the equation of time. Four times in the year the equation is zero, *i.e.*, there is no difference between the clock and the sun. These times are April 15, June 14, Sept. 1, and Dec. 24. At four other times of the year the difference is at a maximum, viz. Feb. 11, sun 14 min. 27 sec. behind clock; May 14, sun 3 min. 49 sec. ahead of clock; July 26, sun 6 min. 16 sec. behind clock; Nov. 2, sun 16 min. 18 sec. ahead of clock. These dates and differences vary very slightly from year to year.

But, whatever measures of time we may use, it is observation of the stars that furnishes the means of correcting them.

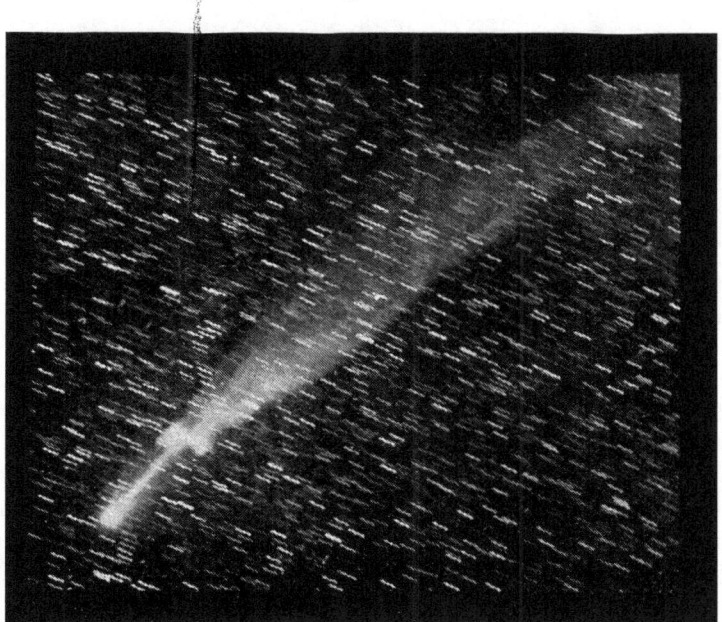

Morehouse's Comet, October 15, 1908

Photographed at the Yerkes Observatory by E. E. Barnard with the ten-inch Bruce telescope. Exposure one hour and a half.

Note the detached portions which appeared to separate from the head and retreat up the line of the tail at enormous velocity.

Morehouse's Comet, November 15, 1908

Photographed at the Yerkes Observatory by E. E. Barnard, with the ten-inch Bruce telescope. Exposure forty minutes.

8. Day and Night. The period of twenty-four hours required for one turn of the earth on its axis is called a day, and in astronomical reckoning it is treated as an undivided whole, the hours being counted uninterruptedly from 0 to 24; but nature has divided the period into two very distinct portions, one characterised by the presence and the other by the absence of the sun. Popularly we speak of the sunlighted portion as day and of the other as night, and there are no two associated phenomena in nature more completely in contrast one to the other. The cause of the contrast between day and night must have been evident to the earliest human beings who were capable of any thought at all. They saw that day inevitably began whenever the sun rose above the horizon, and as inevitably ceased whenever it sank beneath it. In all literatures, imaginative writers have pictured the despair of primeval man when he first saw the sun disappear and night come on, and his joy when he first beheld the sun rise, bringing day back with it. Even his uninstructed mind could not have been in doubt about the causal connection of the sun with daylight.

We now know that the cause of the alternate rising and setting of the sun, and of its apparent motion through the sky, is the rotation of the earth. Making in our minds a picture of the earth as a turning globe exposed to the sunbeams, we are able to see that one half of it must necessarily be illuminated, while the other half is in darkness. We also see that its rotation causes these two halves gradually to interchange places so that daylight progresses completely round the earth once in the course of twenty-four hours. If the earth were not surrounded by an atmosphere, exactly one half of it would lie in the sunlight and exactly one half in darkness, but the atmosphere causes the illuminated part slightly to exceed the unilluminated part. The reason for this is twofold: first, because the atmosphere, being transparent and extending to a considerable height above the solid globe, receives rays from the sun after the latter has sunk below the horizon, and these rays cause a faint illumination in the sky after the sun as viewed from the surface of the ground has disappeared; and, second, because the air has the property of refracting the rays of light, in consequence of which the sun appears above the horizon both a little time before it has actually risen and a little time after it has actually set. The faint illumination at the beginning and the end of the day is called twilight. Its cause is the reflection of light from the air at a considerable elevation above the ground. Observation shows that evening twilight lasts until the sun has sunk about 18° below the western horizon, while morning twilight begins when the sun is still 18° below the nearest horizon. The length of time occupied by twilight, or its duration, depends upon the observer's place on the earth and increases with distance from the equator. The length of twilight at any particular place also varies with the seasons.

It will probably have occurred to the reader that, since day and night are ceaselessly chasing each other round the globe, it must be necessary to choose some point of beginning, in order to keep the regular succession of the days of the week. The necessity for this is evident as soon as we reflect that what is sunrise at one place on the earth, is sunset for a place situated half-way round, on the other side. To understand this it will be better, perhaps, to consider the phenomena of noon at various places. It is noon at any place when the sun is on the meridian of that place. But we have seen that every place has its own meridian; consequently, since the sun cannot be on the meridian of more than one place at a time, each different place (reckoning east and west, for, of course, all places lying exactly north or south of one another have the same meridian), must have its own local noontime. Since the sun appears to move round the earth from east to west, it will arrive at the meridian of a place lying east of us sooner than at our meridian, and it will arrive at our meridian sooner than at that of a place lying west of us. Thus, when it is noon at Greenwich, it is about 7 o'clock A.M., or five hours before noon, at New York, because the angular distance westward round the earth's surface from Greenwich to New York is, in round numbers, 75°, which corresponds with five hours of time, there being 150 to every hour. At the same moment it will be 5 o'clock P.M., or five hours after noon, at Cashmere, because Cashmere lies 75° east of Greenwich. That is to say, the sun crosses the meridian of Cashmere five hours before it reaches the meridian of Greenwich, and it crosses the meridian of Greenwich five hours before it reaches that of New York. At a place half-way round the circumference of the globe, *i.e.* 180° either east or west of Greenwich, it will be midnight at the same instant when it will be mid-day, or noon, at Greenwich. Now let us consider this for a moment.

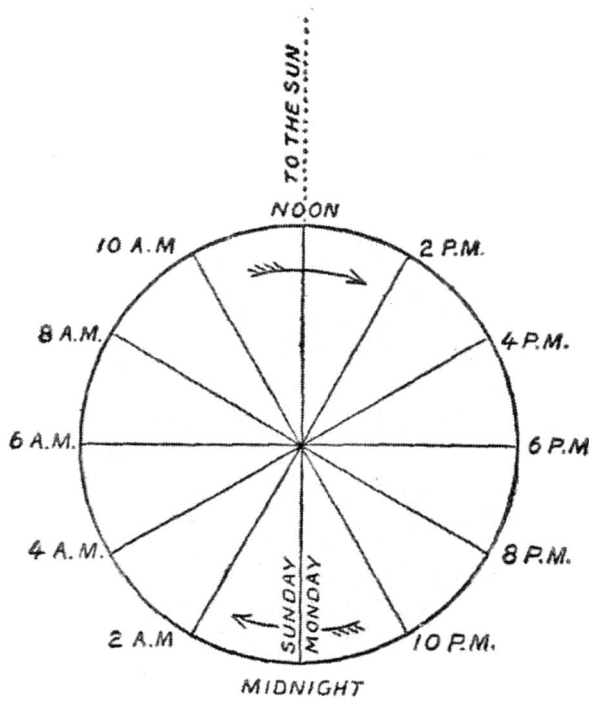

The arrows show the direction in which the earth turns (from west to east). It is always noon at the place which is directly under the sun. Call it Sunday noon at Greenwich, at the top of the circle; then it is 10 A.M. Sunday at a point 30° west and 2 P.M. Sunday at a point 30° east, and so on. Exactly opposite to the noon point it is midnight. By common consent we change the name of the day, and the date, at midnight; consequently it is Sunday midnight just east of the vertical line at the bottom of the circle and Monday morning just west of it. If we cross that line going westward we shall pass directly from Sunday to Monday, and if we cross it going eastward we shall pass directly from Monday to Sunday. Since, by convention, this is a fixed line on the earth's surface, the same change will take place no matter what the hour of the day may be.

It is customary to change the name of the day at midnight. Thus at the stroke of midnight, anywhere, Sunday gives place to Monday. Suppose, then, that the day when we see the sun on the meridian at Greenwich happens to be Sunday. Sunday will then be, so to speak, twelve hours old at Greenwich, because it began there at the preceding midnight. Sunday will be only seven hours old at New York, where it also began at the preceding midnight. In California, 45°, or three hours, still farther west than New York, Sunday will be only four hours old, since the local time there is only four hours after midnight. Go on over the Pacific Ocean, until we arrive at a point 180°, or twelve hours, west of Greenwich. There, evidently, Sunday will just have been born, the preceding day, Saturday, having expired at the stroke of midnight. Now if we just step over that line of 180° in what day shall we be? It cannot be Sunday, because Sunday has just begun on the line itself. It cannot be Saturday, because that would be counting backward. Evidently it can be no other than Monday. Let us examine this a little more closely. It is Sunday noon at Greenwich. We now go round the earth eastward instead of westward. At 90°, or six hours, east of Greenwich, we find that it is 6 P.M. Sunday and at 180°, or twelve hours, east of Greenwich we find that it is Sunday midnight, or in other words Monday morning. But the line of 180° *east* of Greenwich coincides with the line of 180° *west* of Greenwich, which we formerly approached from the opposite direction. So we see that we were right in concluding that in stepping over that line from the east to the west side, we were passing from Sunday into Monday. It is on that line that each day vanishes and its successor takes its place. It is the "date-line" for the whole earth, chosen by the common consent of every civilised nation, just as we have seen that the meridian of Greenwich is the common reference line for reckoning longitude. It lies entirely in the Pacific Ocean, hardly touching any island, and it was chosen for this very reason, because if it ran over inhabited lands, like Europe or America, it would cause endless confusion. Situated as it is, it causes no trouble except to sea captains, and very little to them. If a ship crosses the line going westward the captain jumps his log-book one day forward. If it is, for instance, Wednesday noon, east of the line he calls it Thursday noon, as soon as he has passed over. If he is going eastward he drops back a day on crossing the line, as from Thursday noon to Wednesday noon. The date-line theoretically follows the 180th meridian, but, in fact, in order to avoid certain groups of islands, it bends about a little, while keeping its general direction from north to south.

9. The Seasons. We now recall again what was said in Part I, about the inclination of the ecliptic, or the apparent path of the sun in the heavens, to the equator. Because of this inclination, the sun appears half the year above the equator and the other half below it. When it is above the equator for people living in the northern hemisphere, it is below the equator for those living in the southern hemisphere, and *vice versa.* This is because observers on opposite sides of the plane of the equator look at it from opposite points of view. For the northern observer the celestial equator appears south of the zenith; for the southern observer it appears north of the zenith, its distance from the zenith, in both cases, increasing with the observer's distance from the equator of the earth. If he is on the earth's equator, the celestial equator passes directly *through* the zenith. For convenience we shall suppose the observer to be somewhere in the northern hemisphere.

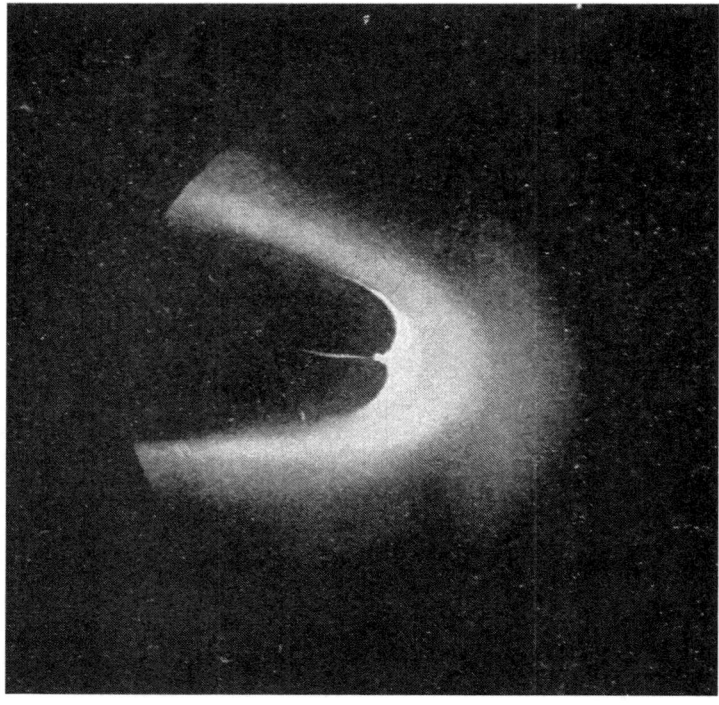

Head of the Great Comet of 1861
From a drawing by Warren De La Rue.

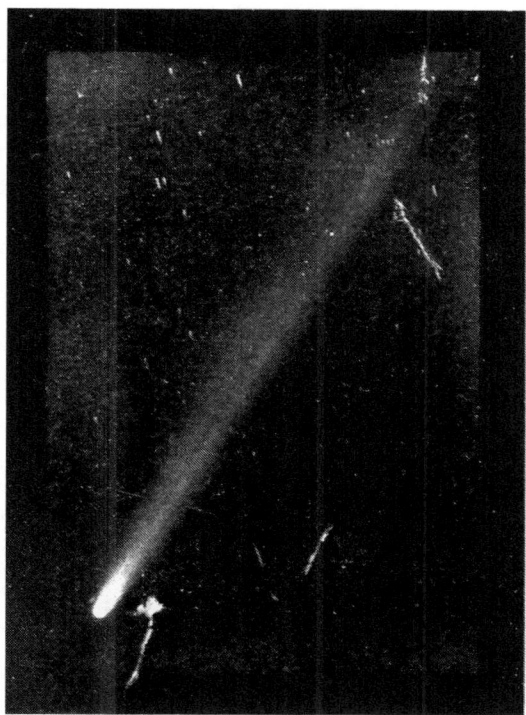

Halley's Comet, May 5, 1910

Photographed at the Yerkes Observatory by E E. Barnard, with the ten-inch Bruce telescope.

This was shortly before the passage of the comet between the earth and the sun, when some think its tail was thrown over us.

Let us begin with that time of the year when the sun arrives at the vernal equinox. This occurs about the 21st of March. The sun is then perpendicular over the equator, daylight extends, uninterrupted, from pole to pole, and day and night (neglecting the effects of twilight and refraction) are of equal length all over the earth. Everywhere there are about twelve hours of daylight and twelve hours of darkness. This is the beginning of the astronomical spring. As time goes on, the motion of the sun in the ecliptic carries it eastward from the vernal equinox, and, at the same time, owing to the inclination of the ecliptic, it rises gradually higher above the equator, increasing its northern declination slowly, day after day. Immediately the equality of day and night ceases, and in the northern hemisphere the day becomes gradually longer in duration than the night, while in the southern hemisphere it becomes shorter. Moreover, because the sun is now north of the equator, daylight no longer extends from pole to pole on the earth, but the south pole is in continual darkness, while the north pole is illuminated.

You can illustrate this, and explain to yourself why the relative length of day and night changes, and why the sun leaves one pole in darkness while rising higher over the other, by suspending a small terrestrial globe with its axis inclined about 23½° from the perpendicular, and passing a lamp around it in a horizontal plane. At two points only in its circuit will the lamp be directly over the equator of the globe. Call one of these points the vernal equinox. You will then see that, when the lamp is directly over this point, its light illuminates the globe from pole to pole, but when it has passed round so as to be at a point higher than the equator, its light no longer reaches the lower pole, although it passes over the upper one.

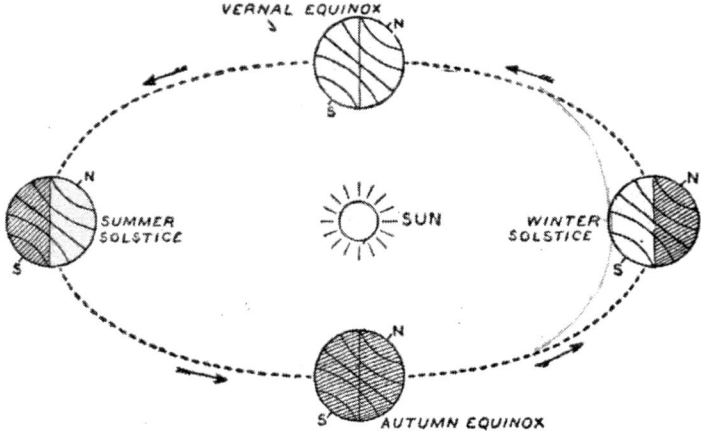

Fig. 11. The Seasons.

The earth is represented at four successive points in its orbit about the sun. Since the axis of the earth is virtually unchangeable in its direction in space (leaving out of account the slow effects of the precession of the equinoxes), it results that at one time of the year, the north pole is inclined toward the sun and at the opposite time of the year away from it. It attains its greatest inclination sunward at the summer solstice, then the line between day and night lies 23½° beyond the north pole, so that the whole area within the arctic circle is in perpetual daylight. The days are longer than the nights throughout the northern hemisphere, but the day becomes longer in proportion to the night as the arctic circle is approached, and beyond that the sun is continually above the horizon. In the southern hemisphere exactly the reverse occurs. When the earth has advanced to the autumn equinox, the axis is inclined neither toward nor away from the sun. The latter is then perpendicular over the equator and day and night are of equal length all over the earth. When the earth reaches the winter solstice the north pole is inclined away from the sun, and now it is summer in the southern hemisphere. At the vernal equinox again there is no inclination of the axis either toward or away from the sun, and once more day and night are everywhere equal. A little study of this diagram will show why on the equator day and night are always of equal length.

Now, with the lamp thus elevated above the equator, set the globe in rotation about its axis. You will perceive that all points in the upper hemisphere are longer in light than in darkness, because the plane dividing the illuminated and the unilluminated halves of the globe is inclined to the globe's axis in such a way that it lies beyond the upper pole as seen from the direction of the lamp. Consequently, the upper half of the globe above the equator, as it goes round, has more of its surface illuminated than unilluminated, and, as it turns on its axis, any point in that upper half, moving round parallel to the equator, is longer in light than in darkness. You will also observe that the ratio of length of the light to the darkness is greater the nearer the point lies to the pole, and that when it is within a certain distance of the pole, corresponding with the elevation of the lamp above the equator, it lies in continual light—in other words, within that distance from the pole night vanishes and daylight is unceasing. At the same time you will perceive that round the lower pole there is a similar space within which day has vanished and night is unceasing, and that in the whole of the lower hemisphere night is longer than day. Exactly on the equator, day and night are always of equal length.

Endeavour to represent all this clearly to your imagination, before actually trying the experiment, or consulting a diagram. If you try the experiment you may, instead of setting the axis of the globe at a slant, place it upright, and then gradually raise and lower the lamp as it is carried round the globe, now above and now below the equator.

We return to our description of the actual movements of the sun. As it rises higher from the equator, not only does the day increase in length relatively to the night, but the rays of sunlight descend more nearly perpendicular upon the northern hemisphere. The consequence is that their heating effect upon the ground and the atmosphere increases and the temperature rises until, when the sun reaches its greatest northern declination, about the 22d of June (when it is 23½° north of the equator), the astronomical summer begins. This point in the sun's course through the circle of the ecliptic is called the summer solstice (see Part I, Sect. 8). Having passed the solstice, the sun begins to decline again toward the equator. For a short time the declination diminishes slowly because the course of the ecliptic close to the solstice is nearly parallel to the equator, and in the meantime the temperature in the northern hemisphere continues to increase, the amount of heat radiated away during the night being less than that received from the sun during the day. This condition continues for about six weeks, the greatest heats of summer falling at the end of July or the beginning of August, when the sun has already declined far toward the equator, and the nights have begun notably to lengthen. But the accumulation of heat during the earlier part of the summer is sufficient to counterbalance the loss caused by the declension of the sun.

About the 23d of September the sun again crosses the equator, this time at the autumnal equinox, the beginning of the astronomical autumn, and after that it sinks lower and lower (while appearing to rise in the southern hemisphere), until about the 22d of December, when it reaches its greatest southern declination, 23½°, at the winter solstice, which marks the beginning of the astronomical winter. It is hardly necessary to point out that the southern winter corresponds in time with the northern summer, and *vice versa*. From the winter solstice the sun turns northward once more, reaching the vernal equinox again on the 21st of March.

Thus we see that we owe the succession of the seasons entirely to the inclination of the earth's axis out of a perpendicular to the plane of the ecliptic. If there were no such inclination there would be climate but no seasons. There would be no summer heat, except in the neighbourhood of the equator, while the middle latitudes would have a moderate temperature the year round. Owing to the effects of refraction, perpetual day would prevail within a small region round each of the poles. The sun would be always perpendicular over the equator.

Two things remain to be pointed out with regard to the effect of the sun's annual motion in the ecliptic. One of these is the circles called the tropics. These are drawn round the earth parallel to the equator and at a distance of 23½° from it, one in the northern and the other in the southern hemisphere. The northern one is called the tropic of Cancer, because its corresponding circle on the celestial sphere runs through the zodiacal sign Cancer, and the southern one is called the tropic of Capricorn for a similar reason. The tropics run through the two solstices, and mark the apparent *daily* track of the sun in the sky when it is at either its greatest northern or its greatest southern declination. The sun is then perpendicular over one or the other of the tropics. That part of the earth lying between the tropics is called the torrid zone, because the sun is always not far from perpendicular over it, and the heat is very great.

The Six-Tailed Comet of 1744
From a contemporary drawing.

The other thing to be mentioned is the polar circles. These are situated 23½° from each pole, just as the tropics are situated a similar distance on each side of the equator. The northern is called the arctic, and the southern the antarctic circle. Those parts of the earth which lie between the tropics and the polar circles are called respectively the northern and the southern temperate zone. The polar circles mark the limits of the region round each pole where the sun shines continuously when it is at one or the other of the solstices. If the reader will recall the experiment with the globe and the lamp, he will perceive that these circles correspond with the borders of the circular spaces at each pole of the globe which are alternately carried into and out of the full light as the lamp is elevated to its greatest height above the equator or depressed to its greatest distance below it. At each pole, in turn, there are six months of continual day followed by six months of continual night, and when the sun is at one of the solstices it just touches the horizon on the corresponding polar circle at the hour that marks midnight on the parts of the earth which lie outside the polar circles. This is the celebrated phenomenon of the "midnight sun." At any point within the polar circle concerned, the sun, at the hour of midnight approaches the horizon but does not touch it, its midnight elevation increasing with nearness to the pole, while exactly at the pole itself the sun simply moves round the sky once in twenty-four hours in a circle practically parallel to the horizon. It is by observations on the daily movement of the sun that an explorer seeking one of the earth's poles during the long polar day is able to determine when he has actually reached his goal.

The reader will have remarked in these descriptions how frequently the angle of 23½° turns up, and he should remember that it is, in every case, due to the same cause, viz., the inclination of the earth's axis from a perpendicular to the ecliptic.

A very remarkable fact must now be referred to. Although the *angular distance* that the sun has to travel in passing first from the vernal equinox to the autumnal equinox, on the northern side of the equator, and then back again from the autumnal equinox to the vernal equinox, on the southern side of the equator, is the same, the *time* that it occupies in making these two half stages in its annual journey is *not* the same. Beginning from the 21st of March and counting the number of days to the 23d of September, and then beginning from the 23d of September and counting the number of days to the next 21st of March, you will find that in an ordinary year the first period is seven days longer than the second. In other words, the sun is a week longer above the equator than below it. The reason for this difference is found in the fact that the orbit of the earth about the sun is not a perfect circle, but is a slightly elongated ellipse, and the sun, instead of being situated in the centre, is situated in one of the two foci of the ellipse, 3,000,000 miles nearer to one end of it than to the other. Now this elliptical orbit of the earth is so situated that the earth is nearest to the focus occupied by the sun, or in perihelion, about December 31st, only a few days after the winter solstice, and farthest from the sun, or in aphelion, about July 1st, only a few days after the summer solstice. Thus the earth is nearer the sun during the winter half of the year, when the sun appears south of the equator, than during the summer half of the year, when the sun appears north of the equator. Now the law of gravitation teaches that when the earth is nearer the sun it must move more rapidly in its orbit than when it is more distant, from which it follows that the time occupied by the sun in its apparent passage from the vernal equinox to the autumnal equinox is longer than that occupied in the passage back from the autumnal to the vernal equinox.

But while the summer half of the year is longer than the winter half in the northern hemisphere, the reverse is the case in the southern hemisphere. There the winter is longer than the summer. Moreover, the winter of the southern hemisphere occurs when the earth is farthest from the sun, which accentuates the disadvantage. It has been thought that the greater quantity of ice about the south pole may be due to this increased length and severity of the southern winter. It is true that the southern summer, although shorter, is hotter than the northern, but while, theoretically, this should restore the balance as a whole, yet it would appear that the short hot summer does not, in fact, suffice to arrest the accumulation of ice.

However, the present condition of things as between the two hemispheres will not continue, but in the course of time will be reversed. The reader will recall that the precession of the equinoxes causes the axis of the earth to turn slowly round in space. At present the northern end of the earth's axis is inclined away from the aphelion and in the direction of the perihelion point of the orbit, so that the northern summer occurs when the earth is in the more distant part of its orbit, and the winter when it is in the nearer part. But the precession swings the axis round *westward* from its present position at the rate of 50 .2 per year, while at the same time the position of the orbit itself is shifted (by the effects of the attraction of the planets) in such a manner that the aphelion and perihelion points, which are called the apsides, move round *eastward* at the rate of 11 .25 per year. The combination of the precession with the motion of the apsides produces a revolution at the rate of 61 .45 per year, which in the course of 10,500 years will completely reverse the existing inclination of the axis with regard to the major diameter of the orbit, so that then the northern hemisphere will have its summer when the earth is near perihelion and its winter when it is near aphelion. The winter, then, will, for us, be long and severe and the summer short though hot.

It has been thought possible that such a state of things may cause, in our hemisphere, a partial renewal of what is known in geology as a glacial period. A glacial period in the southern hemisphere would probably always be less severe than in the northern, because of the great preponderance of sea over land in the southern half of the globe. An ocean climate is more equable than a land climate.

10. The Year, the Calendar, and theMonth. A year is the period of time required for the earth to make one revolution in its orbit about the sun. But, as there are three kinds, or measures, of time, so there are three kinds, or measures, of the year. The first of these is called the sidereal year, but although, like sidereal time, it measures the true length of the period in question, it is not suitable for ordinary use. To understand what is meant by a sidereal year, imagine yourself to be looking at the earth from the sun, and suppose that at some instant you should see the earth exactly in conjunction with a star. When, having gone round the sun, it had come back again to conjunction with the same star, precisely one revolution would have been performed in its orbit, and the period elapsed would be a sidereal year. Practically, the length of the sidereal year is determined by observing when the sun, in its apparent annual journey round the sky, has come back to conjunction with some given star.

The second kind of year is called the tropical year, and it is measured by the period taken by the sun to pass round the sky from one conjunction with the vernal equinox to the next. This period differs slightly from the first, because, owing to the precession of the equinoxes, the vernal equinox is slowly shifting westward, as if to meet the sun in its annual course, from which it results that the sun overtakes the equinox a little before it has completed a sidereal year. The tropical year is about twenty minutes shorter than the sidereal year. It is, however, more convenient for ordinary purposes, because we naturally refer the progress of the year to that of the seasons, and, as we have seen, the seasons depend upon the equinoxes.

But yet the tropical year is not entirely satisfactory as a measure of time, because the number of days contained in it is not an even one. Its length is 366 days, 5 hours, 48 minutes, 46 seconds. Accordingly, as the irregularities of apparent solar time were avoided by the invention of mean solar time, so the difficulty presented by the tropical year is gotten rid of, as far as possible, by means of what is called the civil year, or the calendar year, the average length of which is almost exactly equal to that of the tropical year. This brings us to the consideration of the calendar, which is as full of compromises as a political treaty—but there is no help for it since nature did not see fit to make the day an exact fraction of the year, or, in other words, to make the day and the year commensurable quantities of time.

Without going into a history of the reforms that the calendar has undergone, which would demand a great deal of space, we may simply say that the basis of the calendar we use to-day was established by Julius Cæsar, with the aid of the Greek astronomer Sosigenes. This is the Julian calendar, and the reformed shape in which it exists at present is called the Gregorian calendar. Cæsar assumed 365¼ days as the true length of the year, and, in order to get rid of the quarter day, ordered that it should be left out of account for three years out of every four. In the fourth year the four quarter days were added together to make one additional day, which was added to that particular year. Thus the ordinary years were each 365 days long and every fourth year was 366 days long. This fourth year was called the bissextile year. It was identical with our leap year. The days of both the ordinary and the leap years were distributed among the twelve months very much as we distribute them now.

But Cæsar's assumption of 365¼ days as the length of the year was erroneous, being about 11 min. 14 sec. longer than the real tropical year. In the sixteenth century this error had accumulated to such a degree that the months were becoming seriously disjointed from the seasons with which they had been customarily associated. In consequence, Pope Gregory XIII, assisted by the astronomer Clavius, introduced a slight reform of the Julian calendar. The accumulated days were dropped, and a new start taken, and the rule for leap year was changed so as to read that "all years, whose date-number is divisible by four without a remainder are leap years, unless they are century years (such as 1800, 1900, etc.). The century years are not leap years, unless their date number is divisible by 400, in which case they are." And this is the rule as it prevails to-day, although there is now (1912) serious talk of undertaking a new revision. But the Gregorian calendar is so nearly correct that more than 3000 years must elapse before the length of the year as determined by it will differ by one day from the true tropical year.

The subject of the reform of the calendar is a very interesting one, but, together with that of the rules for determining the date of Easter, its discussion must be sought in more extensive works.

There is one other measure of time, depending upon the motion of a heavenly body, which must be mentioned. This is the month, or the period required for the moon to make a revolution round the earth.   Here we encounter again the same difficulty, for the month also is incommensurable with the year.   Then, too, the length of the month varies according to the way in which it is reckoned.   We have, first, a sidereal revolution of the moon, which is measured by the time taken to pass round the earth from one conjunction with a star to the next.   This is, on the average, 27 days, 7 hours, 43 minutes, 12 seconds.   Next we have a synodical revolution of the moon, which is measured by the time it takes in passing from the phase of new moon round to the same phase again.   This seems the most natural measure of a month, because the changing phases of the moon are its most conspicuous peculiarity.   (These will be explained in Part III.) The length of the month, as thus measured, is, on the average, 29 days, 12 hours, 44 minutes, 3 seconds. The reason why the synodical month is so much longer than the sidereal month is because new moon can occur only when the moon is in conjunction with the sun, *i.e.* exactly between the earth and the sun, and in the interval between two new moons the sun moves onward, so that for the second conjunction the moon must go farther to overtake the sun. It will be observed that both of the month measures are given in average figures. This is because the moon's motion is not quite regular, owing partly to the eccentricity of its orbit and partly to the disturbing effects of the sun's attraction. The length of the sidereal revolution varies to the extent of three hours, and that of the synodical revolution to the extent of thirteen hours.

But, whichever measure of the month we take, it is incommensurate with the year, *i.e.* there is not an even number of months in a year.   In ancient times ceaseless efforts were made to adjust the months to the measure of the year, but we have practically given up the attempt, and in our calendar the lunar months shift along as they will, while the ordinary months are periods of a certain number of days, having no relation to the movements of the moon.

It has been thought that the period called a week, which has been used from time immemorial, may have originated from the fact that the interval from new moon to the first quarter and from first quarter to full moon, etc., is very nearly seven days.   But the week is as incorrigible as all its sisters in the discordant family of time, and there is no more difficult problem for human ingenuity than that of inventing a system of reckoning, in which the days, the weeks, the months, and the years shall be adjusted to the closest possible harmony.

Spiral Nebula in Ursa Major (M 101)

Photographed at the Lick Observatory by J. E. Keeler, with the Crossley reflector. Exposure four hours.

Note the appearance of swift revolution, as if the nebula were throwing itself to pieces like a spinning pin-wheel.

The Whirlpool Nebula in Canes Venatici

Photographed at the Lick Observatory by J. E. Keeler, with the Crossley reflector. Exposure four hours.

Note the "beading" of the arms of the whirling nebula.

PART III. THE SOLAR SYSTEM.

PART III.   THE SOLAR SYSTEM.

1. The Sun. By the term solar system is meant the sun together with the system of bodies (planets, asteroids, comets and meteors) revolving round it. The sun, being a star, every other star, for all that we can tell, may be the ruler of a similar system. In fact, we *know* that a few stars have huge dark bodies revolving round them, which may be likened to gigantic planets. The reason why the sun is the common centre round which the other members of the solar system move, is because it vastly exceeds all of them put together in mass, or quantity of matter, and the power of any body to set another body in motion by its attractive force depends upon mass. If a great body and a small body attract each other, both will move, but the motion of the small body will be so much more than that of the great one that the latter will seem, relatively, to stand fast while the small one moves. Then, if the small body had originally a motion across the direction in which the great body attracts it, the result of the combination will be to cause the small body to revolve in an orbit (more or less elliptical according to the direction and velocity of its original motion) about the great body. If the difference of mass is very great, the large body will remain virtually immovable. This is the case with the sun and its planets. The sun has 332,000 times as much mass (or, we may say, is 332,000 times as heavy) as the earth. It has a little more than a thousand times as much mass as its largest planet, Jupiter. It has millions of times as much as the greatest comet. The consequence is that all of these bodies revolve around the sun, in orbits of various degrees of eccentricity, while the sun itself remains practically immovable, or just swaying a little this way and that, like a huntsman holding his dogs in leash.

The distance of the sun from the earth—about 93,000,000 miles—has been determined by methods which will be briefly explained in the next section. Knowing its distance, it is easy to calculate its size, since the apparent diameter of all objects varies directly with their distance. The diameter of the sun is thus found to be about 866,400 miles, or nearly 110 times that of the earth. In bulk it exceeds the earth about 1,300,000 times, but its mass, or quantity of matter, is only 332,000 times the earth's, because its average density is but one quarter that of the earth. This arises from the fact that the earth is a solid, compact body, while the sun is a body composed of gases and vapours (though in a very compressed state). It is the high temperature of the sun which maintains it in this state. Its temperature has been calculated at about 16,000° Fahrenheit, but various estimates differ rather widely. At any rate, it is so hot that the most refractory substances known to us would be reduced to the state of vapour, if removed to the sun. The quantity of heat received upon the earth from the sun can only be expressed in terms of the mechanical equivalent of heat. The unit of heat in engineering is the calorie, which means the amount of heat required to raise the temperature of one kilogram of water (2.2 pounds) one degree Centigrade (1°.8 Fahrenheit). Now observation shows that the sun furnishes 30 of these calories per minute upon every square metre (about 1.2 square yard) of the earth's surface. Perhaps there is no better illustration of what this means than Prof. Young's statement, that "the heat annually received on each square foot of the earth's surface, if employed in a perfect heat engine, would be able to hoist about a hundred tons to the height of a mile." Or take Prof. Todd's illustration of the mechanical power of the sunbeams: "If we measure off a space five feet square, the energy of the sun's rays when falling vertically upon it is equivalent to one horse power." Astronomers ordinarily reckon the solar constant in "small calories," which are but the thousandth part of the engineer's calorie, and the latest results of the Smithsonian Institution observations indicate that the solar constant is about 1.95 of these small calories per square centimeter per second. About 30 per cent. must be deducted for atmospheric absorption.

Heat, like gravitation and like light, varies inversely in intensity with the square of the distance; hence, if the earth were twice as near as it is to the sun it would receive four times as much heat and four times as much light, and if it were twice as far away it would receive only one quarter as much. This shows how important it is for a planet not to be too near, or too far from, the sun. The earth would be vapourised if it were carried within a quarter of a million miles of the sun.

The sun rotates on an axis inclined about 7½° from a perpendicular to the plane of the ecliptic. The average period of its rotation is about 25 days—we say "average" because, not being a solid body, different parts of its surface turn at different rates. It rotates faster at the equator than at latitudes north-and-south of the equator, the velocity decreasing toward the poles. The period of rotation at the equator is about 25 days, and at 40° north or south of the equator it is about 27 days. The direction of rotation is the same as that of the earth's.

The surface of the sun, when viewed with a telescope, is often seen more or less spotted. The spots are black, or dusky, and frequently of very irregular shapes, although many of them are nearly circular. Generally they appear in groups drawn out in the direction of the solar rotation. Some of these groups cover areas of many millions of square miles, although the sun is so immense that even then they appear to the naked eye (guarded by a dark glass) only as small dark spots on its surface. The centres of sun-spots, are the darkest parts. Generally around the borders of the spots the surface seems to be more or less heaped up. Often, in large sun-spots, immense promontories, very brilliant, project over the dark interior, and many of these are prolonged into bridges of light, apparently traversing the chasms beneath. Constant changes of shape and arrangement take place, and there are few more astonishing telescopic objects than a great sun-spot.

"Tress Nebula" (N. G. C. 6992) in Cygnus

Photographed at the Yerkes Observatory by G. W. Ritchey, with the two-foot reflector.

Observe the strangely twisted look of this long curved nebula; also the curious curves composed of minute stars near it.

The spots are not always visible in equal numbers, and in some years but few are seen, and they are small. It has been found that they occur in periods, averaging about eleven years from maximum to minimum, although the length of the period is very irregular. It has also been observed that when the first spots of a new period appear, they are generally seen some 30° from the equator, either toward the north or toward the south, and that as the period progresses the spots increase in size, and seem to draw toward the equator, the last spots of the period being seen quite close to the equator, on one side or the other. The duration of individual spots is variable; some last but a day or two, and others continue for weeks, sometimes being carried out of view by the rotation of the sun and brought into view again from the other side.

The surface of the sun in the neighbourhood of groups of spots is frequently marked by large areas covered with crinkled bright lines and patches, which are called faculæ. These, which are the brightest parts of the sun, appear to be elevated above the general level.

As to the cause and nature of sun-spots much remains to be learned. In 1908, Prof. George E. Hale, by means of an instrument called the spectro-heliograph, which selects out of the total radiation of the solar disk light peculiar to certain elements, and thus permits the use of that light alone in photographing the sun, demonstrated that sun-spots probably arise from vortices, or whirling storms, and that these vortices produce strong magnetic fields in the sun-spots. The phenomenon may be regarded, says Prof. Hale, as somewhat analogous to a tornado or waterspout on the earth. The whirling trombe becomes wider at the top, carrying the gases from below upward. At the centre of the storm the rapid rotation produces an expansion which cools the gases and causes the appearance of a comparatively dark cloud, which we see as the sun-spot. The vortices whirl in opposite directions on opposite sides of the sun's equator, thus obeying the same law that governs the rotation of cyclones on the earth.

It has long been a question whether the condition of the sun as manifested by the spots upon its surface has an influence upon the meteorology of the earth. It is known that the sun-spot period coincides closely with periodical changes in the earth's magnetism, and great outbursts on the sun have frequently been immediately followed by violent magnetic storms and brilliant displays of the aurora borealis on the earth.

The sun undoubtedly exercises other influences upon the earth than those familiar to us under the names of gravitation, light, and heat; but the nature of these other influences is not yet fully understood.

The brilliant white surface of the sun is called the photosphere. It has been likened to a shell of intensely hot clouds, consisting of substances which are entirely vaporous within the body of the sun. Above the photosphere lies an envelope, estimated to be from 5000 to 10,000 miles thick, known as the chromosphere. It consists mainly of hydrogen and helium, and when seen during a total eclipse, when the globe of the sun is concealed behind the moon, it presents a brilliant scarlet colour. Above this are frequently seen splendid red flame-like objects, named prominences. They are of two varieties—one cloud-like in appearance, and the other resembling spikes, or trees with spreading tops,—but often their forms are infinitely varied. The latter, the so-called eruptive prominences, exhibit rapid motion away from the sun's surface, as if they consisted of matter which has been ejected by explosion. Occasionally these objects have been seen to grow to a height of several hundred thousand miles, with velocities of two or three hundred miles per second.

The sun has still another envelope, of changing form,—the corona. This apparently consists of rare gaseous matter, whose characteristic constituent is an element unknown on the earth, called coronium. The corona appears in the form of a luminous halo, surrounding the hidden sun during a total eclipse, and it often extends outward several million miles. Its shape varies in accordance with the sun-spot period. It has a different appearance and outline at a time of maximum sun-spots from those which it presents at a minimum. There are many things about the corona which suggest the play of electric and magnetic forces. The corona, although evidently always existing, is never seen except during the few minutes of complete obscuration of the sun that occurs in a total eclipse. This is because its light is not sufficiently intense to render it visible, when the atmosphere around the observer is illuminated by the direct rays of sunlight.

2. Parallax. We now return to the question of the sun's distance from the earth, which we treat in a separate section, because thus it is possible to present, at a single view, the entire subject of the measurement of the distances of the heavenly bodies. The common basis of all such measurements is furnished by what is called parallax, which may be defined as the difference of direction of an object when viewed alternately from two separate points. The simplest example of parallax is found in looking at an object first with one eye and then with the other without, in the meantime, altering the position of the head. Suppose you sit in front of a window through which you can see the wall of a house on the opposite side of the street. Choose one of the vertical bars of the window-sash, and, closing the left eye, look at the bar with the right and note where it seems to be projected against the wall. Then close the right eye and open the left, and you will observe that the place of projection of the bar has shifted toward the right. This change of direction is due to parallax and its amount depends both upon the distance between the eyes and upon the distance of the window from the observer. To see how this principle is applied by the astronomer, let us suppose that we wish to ascertain the distance of the moon. The moon is so far away that the distance between the eyes is infinitesimal in comparison, so that no parallactic shift in its direction is apparent on viewing it alternately with the two eyes. But by making the observations from widely separated points on the earth we can produce a parallactic shifting of the moon's position which will be easily measurable.

Let one of the points of observation be in the northern hemisphere and the other in the southern, thousands of miles apart. The two observers might then be compared to the eyes of an enormous head, each of which sees the moon in a measurably different direction. If the northern observer carefully ascertains the angular distance of the moon from his zenith, and the southern observer does the same with regard to *his* zenith, as indicated in Fig. 12, they can, by a combination of their measurements, construct a quadrilateral A C B M, of which all the angles may be ascertained from the two measurements, while the length of the sides A C and B C is already known, since they are each equal to the radius of the earth. With these data it is easy, by the rules of plane trigonometry, to calculate the length of the other sides, and also the length of the straight line from the centre of the earth to the moon. In all such cases the distance between the points of observation is called the base-line, whose length is known to start with, while the angles formed by the lines of direction at the opposite ends of the base-line are ascertained by measurement.

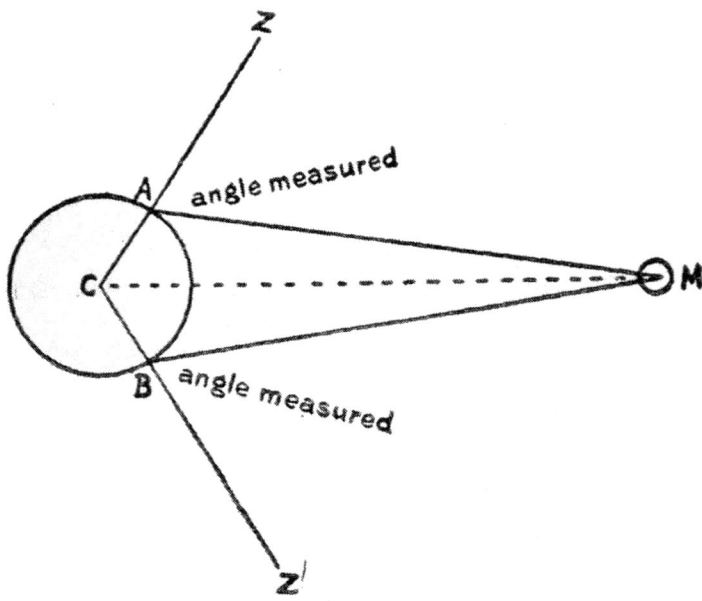

Fig. 12. Parallax of the Moon.

Let C be the centre of the earth, A and B the stations of two observers, one in the northern, the other in the southern hemisphere, and M the moon. The lines C A Z and C B Z indicate the direction of the zenith at A and B respectively. Subtracting the measured angles at A and B each from 180° gives the inside angles at those points. The angle at C is equal to the sum of the latitudes of A and B since they are on opposite sides of the equator. With three angles known, the fourth, at M, is found by simply subtracting their sum from 360°.

The Great Andromeda Nebula

Photographed at the Yerkes Observatory by G. W. Ritchey, with the two-foot reflector.

Observe the vast spiral, or elliptic, rings surrounding the central condensation and the appearance of breaking up and re-shaping into smaller masses which some of the rings present.

In the case of the sun the distance concerned is so great (about 400 times that of the moon) that the parallax produced by viewing it from different points on the earth is too small to be certainly measured, and a modification of the method has to be employed. One such modification, which has been much used, depends upon the fact that the planet Venus, being nearer the sun than the earth is, appears, at certain times, passing directly over the face of the sun. This is called a transit of Venus. During a transit, Venus is between three and four times nearer the earth than the sun is, and consequently its parallactic displacement, when viewed from widely separated points on the earth, is much greater than that of the sun. One of the ways in which the astronomer takes advantage of this fact is shown in Fig. 13. Let A and B be two points on opposite sides of the earth, but both somewhere near the equator. As Venus swings along in its orbit to pass between the earth and the sun, it will manifestly be seen just touching the sun's edge sooner from A than from B. The observer at A notes with extreme accuracy the exact moment when he sees Venus apparently touch the sun. Several minutes later, the observer at B will see the same phenomenon, and he also notes accurately the time of the apparent contact. Now, since we know from ordinary observation the time that Venus requires to make one complete circuit of its orbit, we can, by simple proportion, calculate, from the time that it takes to pass from v to v1, the angular distance between the lines A S and B S, or in other words the size of the angle at S, which is equal to the parallactic displacement of the sun, as seen from opposite ends of the earth's diameter. Knowing, to begin with, the distance between A and B, we have the means of determining the length of all the other lines in the triangle, and hence the distance of the sun. This process is known as Delisle's method. There is another method, called Halley's, but in a brief treatise of this kind we cannot enter into a description of it. It suffices to say that both depend upon the same fundamental principles.

Fig. 13. Parallax of the Sun from Transit of Venus.
(For description see text.)

It must be added, however, that other ways of measuring the sun's distance than are afforded by transits of Venus have been developed. One of these depends upon observation of the asteroid Eros, which periodically approaches much nearer to the earth than Venus ever does. By observing the parallax of Eros, when it is nearest the earth, its distance can be ascertained, and that being known the distance of the sun is immediately deducible from it, because, by the third law of Kepler (to be explained later), the *relative* distances of all the planets from the sun are proportional to their periods of revolution, so that if we know any one of the distances in miles we can calculate all the others. It is important here to state the angular amount of the sun's parallax, since it is a quantity which is continually referred to in books on astronomy. According to the latest determination, based on observations of Eros, the solar parallax is 8 .807, which corresponds, in round numbers, to a distance of 92,800,000 miles. A mean parallax of 8 .796 is given by Mr. C. G. Abbot, based on a combination of results from a number of different methods, and this corresponds to a distance of 92,930,000 miles. To the astronomer, who seeks extreme exactness, the slightest difference is important. It should be noted that the figures 8 .807 or 8 .796 represent the parallactic displacement of the sun, as seen not from the opposite ends of the earth's entire diameter, but from opposite ends of its radius, or semi-diameter. Accordingly it is equal to half of the angle at S in Fig. 13. It is for convenience of calculation that, in such cases, the astronomer employs the semi-diameter, instead of the whole diameter for his base-line.

The case of the stars must next be considered, and now we find that the distances involved are so enormous that the diameter, or semi-diameter, of the earth is altogether too insignificant a quantity to afford an available base-line for the measurement. We should have remained forever ignorant of star distances but for the effects produced by the earth's change of place due to its annual revolution round the sun. The mean diameter of the earth's orbit is about 186,000,000 miles, and we are able to make use of this immense distance as a base-line for ascertaining the parallax of a star. Suppose, for instance, that the direction of a star in the sky is observed on the 1st of January, and again on the 1st of July. In the meantime, the earth will have passed from one end of the base-line just described to the other, and unless the star observed is extremely remote, a careful comparison of the two measurements of direction will reveal a perceptible parallax, from which the actual distance of the star in question can be deduced.

It is to be observed that if all the stars were equally distant this method would fail, because then there would be no "background" against which the shift of place could be observed; all of the stars would shift together. But, in fact, the vast majority of the stars are so remote that even a base-line of 186,000,000 miles is insufficient to produce a measurable shift in their direction. It is only the distances of the nearer stars which we can measure, and for them the multitude of more remote ones serves, like the wall of the house in the experiment with the window-bar, as a background on which the shift of place can be noted. Just as in calculations of the sun's parallax the semi-diameter of the earth is chosen for a base-line, so in the case of the stars the semi-diameter of the earth's orbit, amounting to 93,000,000 miles, forms the basis. Measured in this way the parallaxes of the nearest stars come out in tenths, or hundredths, of a second of arc, or angular measurement. Thus the parallax of Alpha Centauri, the nearest known star, is about 0 .75, corresponding to a distance of about 26,000,000,000,000 miles. Now 0 .75 is a quantity inappreciable to the naked eye, and only to be measured with delicate instruments, and yet this almost invisible shift of direction is all that is produced by viewing the nearest star in the sky from the opposite ends of a base-line 93,000,000 miles long!

3. Spectroscopic Analysis. We have next to deal with the constitution of the sun, or the nature of the substances of which it consists, and for this purpose we must first understand the operation of the spectroscope, in many respects the most wonderful instrument that man has invented. It has given birth to the "chemistry of the sun" and the "chemistry of the stars," for by its aid we can be as certain of the nature of many of the substances of which they are made as we could be by actually visiting them.

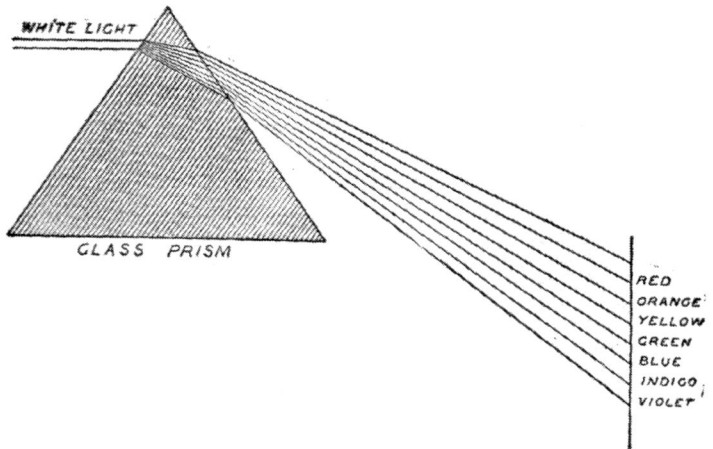

WHITE LIGHT

GLASS PRISM

RED
ORANGE
YELLOW
GREEN
BLUE
INDIGO
VIOLET

Fig. 14. Spectrum Analysis.

The red is least turned by the prism from its original course and the violet most. If between the prism and the screen on which the spectrum falls there were interposed a gas of any kind that gas would absorb from the coloured rays passing through it the exact waves of light with which it would itself shine if it were made luminous by heat. It would not take out an entire section, or colour, from the spectrum, but only a small part of one or more of the colours, and the absence of these parts would be indicated on the screen by narrow black lines situated in various places; and these lines, in number and in situation, would differ with every different kind of gas that was interposed. If several kinds were interposed simultaneously they would all pick out their own peculiar rays from the light, and thus the spectrum would be crossed by a large number of dark lines, by the aid of which the nature of the various gases that produced them could be told. The effect would be the same if the gases were interposed in the path of the white light before it enters the prism;—and this, in fact, is what happens when the spectrum of the sun, or a star, is examined—the absorption has already occurred at the surface of the luminous body before the light comes to the earth.

Fundamentally, spectroscopic analysis depends upon the principle of refraction, of which we have spoken in connection with the atmosphere. Although the most powerful spectroscopes are now made on a different plan, the working of the instrument can best be comprehended by considering it in the form in which it was first invented, and in which it is still most often used. In its simplest form the spectroscope consists of a three-angle prism of glass, through which a ray of light is sent from the sun, star, or other luminous object to be examined. Glass, like air or water, has the property of refracting, or bending, all rays of light that enter it in an inclined direction. In passing through two of the opposite-sloping sides of a prism, the ray is twice bent, once on entering and again on leaving, in accordance with the principle that we have already mentioned (see Part II, Sect. 4). Still, merely bending the ray out of its original course would have no important result but for another associated phenomenon, known as dispersion. To explain dispersion we must recall the familiar fact that white light consists of a number of coloured components which, when united, make white. It is usual to speak of these primary, or prismatic, colours, as seven in number. These are red, orange, yellow, green, blue, indigo, and violet. Physicists now assign a different list of primary colours, but these, being generally familiar, will best serve our purpose. Without going into an explanation of the reasons, it will suffice to say that the waves of light producing these fundamental colours are not all equally affected by refraction. The red is least, and the violet most, bent out of its course in passing through the prism, the other colours being bent more and more in proportion to their distance from the red. It follows that the ray, or beam, of light, which was white when it entered the prism, becomes divided or dispersed during its passage into a brush of seven different hues. Thus the prism may be said to analyse the light into its fundamental colours, making them separately visible. This, as a scientific fact, dates from the time of Newton. But Newton did not dream of the further magic that lay in the prism.

It was noticed as early as 1801 that, when the light of the sun was dispersed in the way we have described, not only did the seven primary colours make their appearance, but across the ribbon-like band, called the spectrum, that was thus formed, ran a number of thin black lines, like narrow gaps in the band. The position of these lines was carefully studied by a German astronomer, Fraunhofer, in 1814, and they still bear the name of Fraunhofer lines. But the full explanation of them did not come until 1858 when, with their aid, Kirschoff laid the foundations of spectrum analysis.

This analysis is based upon the fact that the Fraunhofer lines are visual indications of the existence of certain substances in the sun. To explain this we must know three fundamental facts:

1st: Every incandescent body that is either solid or liquid (or, if it consists of gases, is under high pressure) shines with compound white light, which, when dispersed by prisms, gives a *continuous* coloured band, or spectrum.

2d: Every elementary substance when in the gaseous state, and under low pressure, if brought to incandescence by heat, shines with light which, when dispersed, gives a *discontinuous* spectrum, made up of separate bright lines; and each different element possesses its own peculiar spectral lines, never coinciding in position with the lines of any other element.

3d: If the light from a body giving a continuous spectrum is caused to pass through a gas which is at a lower temperature, the gas will absorb precisely those light waves, of which its own spectrum is composed and will leave in the spectrum of the body a series of dark lines, or gaps, whose number and position indicate the nature of the gas whose absorptive action has produced them.

Now, to apply these principles to the sun we have only to remember that it is a globe of gaseous substances, which are under great pressure, owing to the immense force of the sun's gravitation. Consequently it gives a continuous spectrum. But, at the same time, it is surrounded with gaseous envelopes, which are not as much compressed as the internal gases are, and which are at a lower temperature because they come in contact with the cold of surrounding space. The light from the body of the sun must necessarily pass through these envelopes, and each of the gases of which they consist absorbs from the passing sunlight its own peculiar rays, with the result that the spectrum of the sun is seen crossed with a great number of black lines—the Fraunhofer lines.

It will be remarked that the evidence which the Fraunhofer lines afford concerning the composition of the sun applies, strictly, only to the outer portion, or to the envelopes of gaseous matter that surround the interior globe. But since there is every reason to believe that the entire body of the sun is in a gaseous state, notwithstanding the internal pressure, and since we see that there is a continual circulation going on between the inner and outer portions, it is logical to conclude that essentially the same elements exist under varying conditions in all parts of the sun.

In this way, then, we have learned the composition of the sun, and we find that it consists of virtually the same elementary substances found upon the earth, but existing there in a gaseous or vaporous state. Among the elements which have been positively identified in the sun by means of their characteristic spectral lines are iron, calcium, sodium, aluminum, copper, zinc, silver, lead, potassium, nickel, tin, silicon, manganese, magnesium, cobalt, hydrogen, and at least twenty others which are likewise found upon the earth. Some elementary substances known on the earth, such as gold and oxygen, have not yet been certainly found in the sun, but there is every reason to believe that they all exist there, though perhaps under conditions which render their detection difficult or impossible. Helium was recognised as an element in the sun, by giving spectral lines different from any known substance, and it received its name "sun-metal," long before it was discovered on the earth. We have seen that there is at least one element in the sun, coronium, which, as far as we know, does not exist at all upon the earth, and it is not improbable that there may be others which have no counterparts on the earth.

The same kind of analysis applies to the stars, no matter how far away they may be, so long as they give sufficient light to form a spectrum. And in this way it has been found that the stars differ somewhat from the sun and from one another in their composition, and thus a classification of the stars has been made, and it has been possible to draw conclusions concerning their relative age, which show that some stars are comparatively younger than the sun, others older, and others so far advanced in age, or evolution, that they are drawing near extinction. Many dark bodies also exist among the stars, which appear to be completely extinguished suns. It only remains to add on this subject that, according to prevailing theories, the earth itself was once an incandescent body, shining with its own light, and at that time it, too, would have yielded a spectrum showing of what substances it consisted.

4. The Moon. The earth is a satellite of the sun, and the moon is a satellite of the earth. The mean, or average, distance of the sun from the earth is about 93,000,000 miles; the mean distance of the moon is a little less than 239,000 miles. This distance is variable to the extent of about 31,000 miles, owing to the eccentricity of the moon's orbit about the earth. That is, the moon is sometimes nearly 253,000 miles away, and sometimes only about 221,600. The diameter of the moon is 2163 miles. Its bulk is one-forty-ninth that of the earth, but its mass is only one-eightieth, because its mean density is only about six-tenths as great as the earth's.

The moon appears to travel in an orbit round the earth, but in fact the orbit is always concave toward the sun, and the disturbing attraction of the earth, as the two move together round the sun, causes it to appear now on one side and now on the other. But we may treat the moon's orbit as if the earth were the true centre of force, the attraction of the sun being regarded as the disturbing element.

According to a mathematical theory, which has been largely accepted as probably true, but into which we cannot enter here (see Prof. George Darwin's *The Tides*, or Prof. R. Ball's *Time and Tide*), the moon was thrown off from the earth many ages ago, as a consequence of tidal "friction." As it moves round in its orbit the moon keeps the same face toward the earth. This fact is also ascribed to tidal influence.

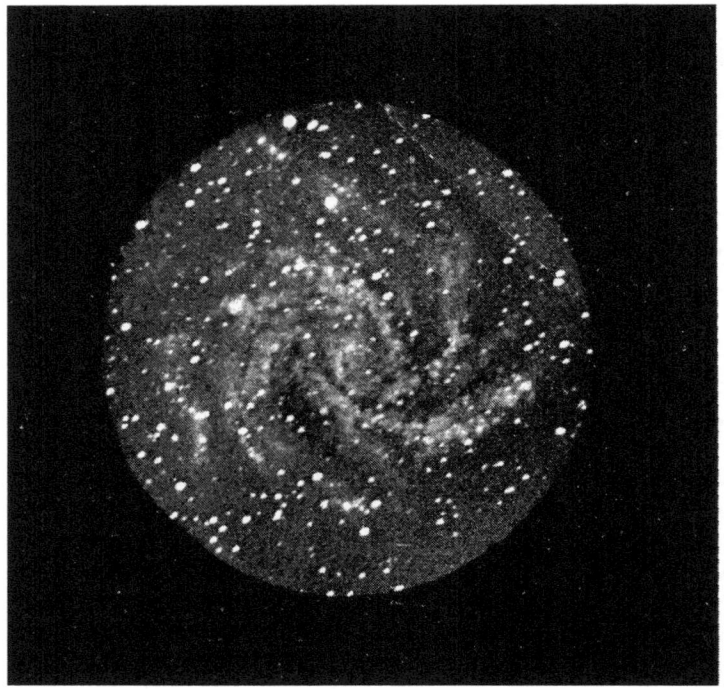

Spiral Nebula in Cepheus (H IV 76)

Photographed at the Lick Observatory by J. E. Keeler, with the Crossley reflector. Exposure four hours.

Observe that the central portion is only of stellar magnitude.

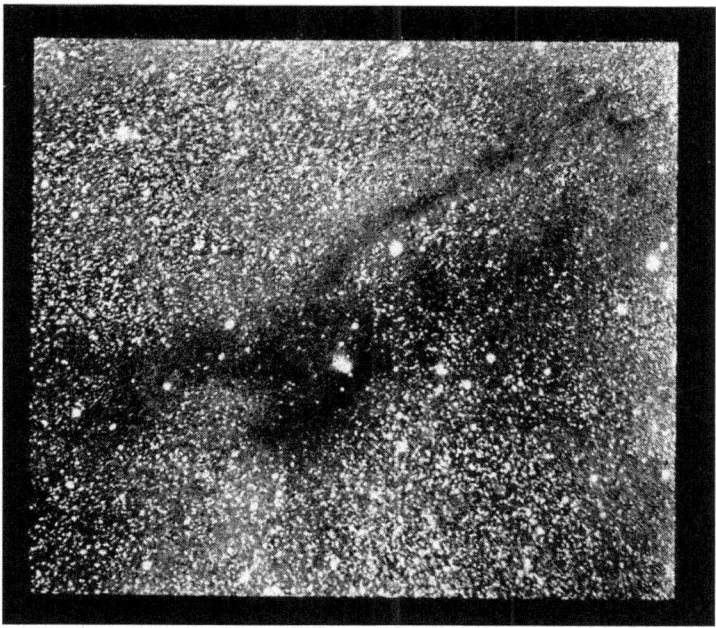

Nebulous Groundwork in Taurus

Photographed at the Yerkes Observatory by E. E. Barnard with 10-inch Bruce telescope. Exposure six hours twenty-eight minutes.

Prof. Barnard has suggested that some of these dark lanes in rich regions of stars are non-luminous nebulæ.

Apparently the moon has no atmosphere, or if it has any it is too rare to be certainly detected. On its surface, there is no appearance of water. Consequently we cannot suppose it to be inhabited, at least by any forms of life familiar to us on the earth. But when the moon is viewed with a telescope large relatively flat areas are seen, which some think may have been the beds of seas in ancient times. They are still called *maria*, or "seas," and are visible to the naked eye in the form of great irregular dusky regions. Nearly two-thirds of the surface of the moon, as we see it, consists of bright regions, which are very broken and mountainous. Most of the mountains of the moon are roughly circular, surrounding enormous depressions, which look like gigantic pits. For this reason they are called lunar volcanoes, but, to say nothing of their immense size—for many are fifty or sixty miles across—they differ in many ways from the volcanoes of the earth. It suffices to point out that what resemble volcanic craters are not situated, as is the case on the earth, at the summits of mountains, but are vast sink-holes, descending thousands of feet below the general surface of the moon. Their real origin is unknown, but it is possible that volcanic forces may have produced them. (For a description, with photographs, of these gigantic formations in the lunar world, see the present author's *The Moon.*) In addition to the circular mountains, or craters, there are several long and lofty chains of lunar mountains much resembling terrestrial mountain ranges.

As to the absence of air and water from the moon, some have supposed that they once existed, but, in the course of ages, have disappeared, either by absorption, partly mechanical and partly chemical, into the interior rocks, or by escaping into space on account of the slight force of gravity on the moon, which appears to be insufficient to enable it to retain, permanently, such volatile gases as oxygen, hydrogen, and nitrogen. This leads us to consider the force of the moon's attraction at its surface. We have seen that spherical bodies attract as if their whole mass were collected at their centres. We also know that the force of attraction varies directly as the mass of the attracting body and inversely as the square of the distance from its centre. Now the mass of the moon is one-eightieth that of the earth, so that, upon bodies situated at an equal distance from the centres of both, the moon's attraction would be only one-eightieth of the earth's. But the diameter of the moon is not very much more than one quarter that of the earth, and for the sake of round numbers let us call it one quarter. It follows that an object on the surface of the moon is four times nearer the centre of attraction than is an object on the surface of the earth, and since the force varies inversely as the square of the distance the moon's attraction upon bodies on its surface is relatively sixteen times as great as the earth's. But the total force of the earth's attraction is eighty times greater than the moon's. In order, then, to find the real relative attraction of the moon at its surface we must divide 80 by 16, the quotient, 5, showing the ratio of the earth's force of attraction at its surface to that of the moon at *its* surface. In other words, this calculation shows that the moon draws bodies on its surface with only one-fifth the force with which the same bodies would be drawn on the earth's surface. The weight of bodies of equal mass would, therefore, be only one-fifth as great on the moon as on the earth.

But the real difference is greater than this, for we have used round numbers, which exaggerated the size of the earth as compared with that of the moon. If we employ the fractional numbers which show the actual ratio of the moon's radius (half-diameter) to that of the earth, we shall find that the weight of the same body would be only about one-sixth as great on the moon as on the earth. It has been thought that this relative lack of weight on the moon may account for the gigantic proportions assumed by its craters, since the same elective force would throw volcanic matter to a much greater height and distance there than on our planet.

The connection of the slight force of gravity on the moon with its ability to retain an atmosphere is shown by the following considerations. It is possible to calculate for any planet of known mass the velocity with which a particle would have to move in order to escape from the control of that planet. In the case of the earth this critical velocity, as it is called, amounts to about 7 miles per second, and in the case of the moon to only 1½ miles per second. Now the kinetic theory of gases informs us that their molecules are continually flying in all directions with velocities varying with the nature and the temperature of the gas. The maximum velocity of the molecules of oxygen is 1.8 miles per second, of hydrogen 7.4 miles, of nitrogen 2 miles, of water vapour 2.5 miles. It is evident, then, that the force of the earth's attraction is sufficient permanently to retain all these gases except hydrogen, and in fact there is no gaseous hydrogen in the atmosphere, that element being found on the earth only in combination with other substances. But oxygen and nitrogen, which constitute the bulk of the atmosphere, have maximum molecular velocities much less than the critical velocity above described. In the case of the moon, however, the critical velocity is less than those of the molecules of oxygen, nitrogen, and water vapour, to say nothing of hydrogen; therefore the moon cannot permanently retain them. We say "permanently," because they might be retained for a time for the reason that the molecules of a gas fly in all directions, and continual collisions occur among them in the interior of the gaseous mass, so that it would be only those at the exterior of the atmosphere which would escape; but gradually all that remained free from combination would get away.

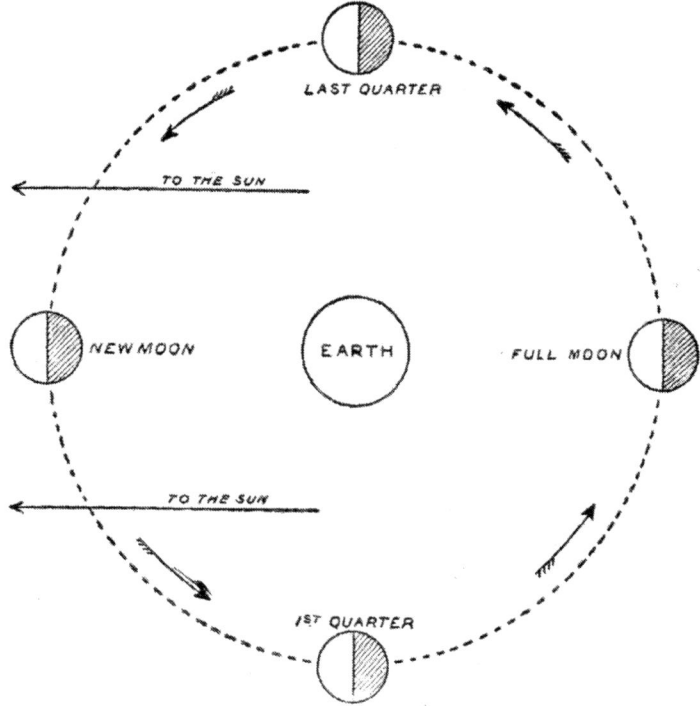

Fig. 15. The Phases of the Moon.

As the moon goes round the earth in the direction indicated by the arrows, the sun remaining always on the left-hand side, it is evident that the illuminated half of the moon will be turned away from the earth at new moon, and toward it at full moon, while between these positions more or less will be seen according to the direction of the moon with regard to the sun.

As the moon travels round the earth it shows itself in different forms, gradually changing from one into another, which are known as phases. If the moon shone with light of its own its outline would always be circular, like the sun's. The apparent change of form is due, first, to its being an opaque globe, reflecting the sunlight that falls upon it, and necessarily illuminated on only one side at a time; and second, to the fact that as it travels round the earth the half illuminated by the sun is sometimes turned directly toward us, at other times only partly toward us, and at still other times directly away from us. When it is in that part of its orbit which passes between the sun and the earth, the moon, so to speak, has its back turned to us, the illuminated side being, of course, toward the sun. It is then invisible, and this unseen phase is the true "new moon." It is customary, however, to give the name new moon to the narrow, sickle-shaped figure, which it shows in the west, after sunset, a few days after the date of the real new moon. The sickle gradually enlarges into a half circle as the moon passes away from the sun, and the half circle phase, which occurs when the moon arrives at a position in the sky at right angles to the direction of the sun, is called first quarter. After first quarter the moon begins to move round behind the earth, with respect to the sun, and when it has arrived just behind the earth, its whole illuminated face is turned toward the earth, because the sun, which causes the illumination, is on that same side. This phase is called full moon. Afterward the moon returns round the other part of its orbit toward its original position between the earth and the sun, and as it does so, it again assumes, first, the form of a half circle, which in this case is called third, or last, quarter, then that of a sickle, known as "old moon," and finally disappears once more to become new moon again.

A perfectly evident explanation of these changes of form, clearer than any description, can be graphically obtained in this way. Take a billiard ball, a croquet ball, or a perfectly round, smooth, and tightly rolled ball of white yarn, and, placing yourself not too near a brightly burning lamp and sitting on a piano stool (in order to turn more easily), hold the ball up in the light, and cause it to revolve round you by turning upon the stool. As it passes from a position between you and the lamp to one on the opposite side from the lamp, and so on round to its original position, you will see its illuminated half go through all the changes of form exhibited by the moon, and you will need no further explanation of the lunar phases.

Nebula in Sagittarius (M 8)

Photographed at the Lick Observatory by J. E. Keeler, with the Crossley reflector. Exposure three hours.

Note the clustering of stars over the whole field, the intricate forms of the nebula, and particularly the curious black spots, or "holes," resembling drops of ink.

The Harvest Moon and the Hunter's Moon, which are popularly celebrated not only on account of their romantic associations, but also because in some parts of the world they afford a useful prolongation of light after sunset, occur only near the time of the autumnal equinox, and they are always full moons. The full moon nearest the date of the equinox, September 23d, is the Harvest Moon, and the full moon next following is the Hunter's Moon. Their peculiarity is that they rise, for several successive evenings, almost at the same hour, immediately after sunset. This is due to the fact that at that time of the year the ecliptic, from which the moon's path does not very widely depart, is, in high latitudes, nearly parallel with the horizon.

The full moon in winter runs higher in the sky, and consequently gives a brighter light, than in summer. The reason is because, since the full moon must always be opposite to the sun, and since toe sun in winter runs low, being south of the equator, the full moon rides proportionally high.

5. Eclipses. We have mentioned the connection of the moon with the tides, but there is another phenomenon in which the moon plays the most conspicuous part—that of eclipses. There are two kinds of eclipses—solar and lunar. In the former it is the moon that causes the eclipse, by hiding the sun from view; and in the latter it is the moon that suffers the eclipse, by passing through the shadow which the earth casts into space on the side away from the sun. In both cases, in order that there may be an eclipse it is necessary that the three bodies, the moon, the sun, and the earth, shall be nearly on a straight line, drawn through their centres. Since the moon occupies about a month in going round the earth there would be two eclipses in every such period (one of the sun and the other of the moon), if the moon's orbit lay exactly in the plane of the ecliptic, or of the earth's orbit. But, in fact, the orbit of the moon is inclined to that plane at an angle of something over 5°. Even so, there would be eclipses every month if the two opposite points, called nodes, where the moon crosses the plane of the ecliptic, always lay in a direct line with the earth and the sun; but they do not lie thus. If, then, the moon comes between the earth and the sun when she is in a part of her orbit several degrees above or below the plane of the ecliptic, it is evident that she will pass either above or below the straight line joining the centres of the earth and the sun, and consequently cannot hide the latter. But, since eclipses do occur in some months and do not occur in others, we must conclude that the situation of the nodes changes, and such is the fact. In consequence of the conflicting attractions of the sun and the earth, the orbit of the moon, although, like that of the earth, it always retains nearly the same shape and the same inclination, swings round in space, so that the nodes, or crossing points on the ecliptic, continually change their position, revolving round the earth. This motion may be compared to that of the precession of the equinoxes, but it is much more rapid, a complete revolution occurring in a period of about nineteen years.

From this it follows that sometimes the moon in passing its nodes will be in a line with the sun, and sometimes will not. But, owing to the fact that the sun and moon are not mere points, but on the contrary present to our view circular disks, each about half a degree in diameter, an eclipse may occur even if the moon is not in an exact line with the centres of the sun and the earth. The edge of the moon will overlap the sun, and there will be a partial eclipse, if the centres of the two bodies are within one degree apart. Now, the inclination of the moon's orbit to the ecliptic being only a little over 5°, it is apparent that in approaching one of its nodes, along so gentle a slope, it will come within less than a degree of the ecliptic while still quite far from the node. Thus, eclipses occur for a considerable time before and after the moon passes a node. The distances on each side of the node, within which an eclipse of the sun may occur, are called the solar ecliptic limits, and they amount to 18° in either direction, or 36° in sum. Within these limits the sun may be wholly or partially eclipsed according as the moon is nearer to, or farther from, the node. If she is exactly at, or very close to, the node the eclipse will be total.

Solar eclipses vary in another way. What would be a total eclipse, under other circumstances, may be only an annular eclipse if the moon happens to be near her greatest distance from the earth. We have described the variations in her distance due to the eccentricity of her orbit, and we have said that the orbit itself swings round the earth in such a way as to cause the nodes continually to change their places on the ecliptic. The motion of the orbit also causes the lunar apsides, or the points where she is at her greatest and least distances from the earth, to change their places, but their revolution is opposite in direction to that of the nodes, as the revolution of the apsides of the earth's orbit is opposite to that of the equinoxes. The moon's apsides sometimes move eastward and sometimes westward, but upon the whole the eastward motion prevails and the apsides complete one revolution in that direction once in about nine and one-half years. In consequence of the combined effects of the revolution of the nodes and that of the apsides, the moon is sometimes at her greatest distance from the earth at the moment when she passes centrally over the sun, and sometimes at her least distance, or she may be at any intervening distance. If she is in the nearer part of her orbit, her disk just covers that of the sun, and the eclipse is total; if she is in the farther part (since the apparent size of bodies diminishes with increase of distance), her disk does not entirely cover the sun, and a rim of the latter is visible all around the moon. This is called an annular eclipse, because of the ring shape of the part of the sun remaining visible.

The length of the shadow which the moon casts toward the earth during a solar eclipse also plays an important part in these phenomena. This length varies with the distance from the sun. Since the moon accompanies the earth, it follows that when the latter is in aphelion, or at its greatest distance from the sun, the moon is also at its greatest mean distance from the sun, and the length of the lunar shadow may, in such circumstances, be as much as 236,000 miles. When the earth, attended by the moon, is in perihelion, the length of the moon's shadow may be only about 228,000 miles. The average length of the shadow is about 232,000 miles. This is nearly 7000 miles less than the average distance of the moon from the earth, so it is evident that generally the shadow is too short to reach the earth, and it would never reach it, and there would never be a total eclipse of the sun, but for the varying distance of the moon from the earth. When the moon is nearest the earth, or in perigee, its distance may be as small as 221,600 miles, and in all cases when near perigee it is near enough for the shadow to reach the earth.

Inasmuch, as the moon's shadow, even under the most favourable circumstances, is diminished almost to a point before touching the earth, it hardly need be said that it can cover but a very small area on the earth's surface. Its greatest possible diameter cannot exceed about 167 miles, but ordinarily it is much smaller. If both the earth and the moon were motionless, this shadow would be a round or oblong dot on the earth, its shape varying according as it fell square or sloping to the surface; but since the moon is continually advancing in its orbit, and the earth is continually rotating on its axis, the shadow moves across the earth, in a general west to east direction. But the precise direction varies with circumstances, as does also the speed. The latter can never be less than about a thousand miles per hour, and that, only in the neighbourhood of the equator. The moon advances eastward about 2100 miles per hour, and the earth's surface turns in the same direction with a velocity diminishing from about a thousand miles an hour at the equator to 0 at the poles. It is the difference between the velocity of the earth's rotation and that of the moon's orbital revolution which determines the speed of the shadow. The greatest time, which the shadow can occupy in passing a particular point on the earth is only eight minutes, but ordinarily this is reduced to one, two, or three minutes. The true shadow only lasts during the time that the moon covers the whole face of the sun, but before and after this total obscuration of the solar disk the sun is seen partially covered by the moon, and these partial phases of the eclipse may be seen from places far aside from the track which the central shadow pursues. It is only during a total eclipse, and only from points situated within the shadow track, that the solar corona is visible.

In a lunar eclipse it is the earth that is the intervening body, and its shadow falls upon the moon. A solar eclipse can only occur at the time of new moon, and a lunar eclipse only at the time of full moon. The shadow of the earth is much longer and broader than that of the moon, and it never fails to reach the moon, so that it is not necessary here to consider its varying length. The width, or diameter, of the shadow at the average distance of the moon from the earth is about 5700 miles. The moon may pass through the centre of the shadow, or to one side of the centre, or merely dip into the edge of it. When it goes deep enough into the shadow to be entirely covered, the eclipse is total; otherwise it is partial. Since in a total lunar eclipse the entire moon is covered by the shadow, it is evident that such an eclipse, unlike a solar one, may be visible simultaneously from all parts of the earth which, at the time, lie on the side facing the moon. In other words, the earth's shadow does not make merely a narrow track across the face of the moon, but completely buries it. When the moon passes centrally through the shadow, she may remain totally obscured for about two hours. But the moon does not completely disappear at such times, because the refraction of the earth's atmosphere bends a little sunlight round its edges and casts it into the shadow. If the atmosphere round the edges of the earth happens to be thickly charged with clouds, but little light is thus refracted into the shadow, and the moon appears very faint, or almost entirely disappears. But this is rare, and ordinarily the eclipsed moon shines with a pale copperish light.

The occurrence of a lunar eclipse is governed by similar circumstances to those affecting solar eclipses. The lunar ecliptic limits, or the distance on each side of the node within which an eclipse may occur, vary from 9½° to 12¼°

in either direction.

Taking all the various circumstances into account, it is found that there may be, though rarely, seven eclipses in a year, two being of the moon and five of the sun, and that the least possible number of eclipses in a year is two, in which case both will be of the sun. Taking into account also all the various positions which the sun and moon occupy with regard to the earth, it is found that there exists a period of 18 years, 11 days, 8 hours, at the return of which eclipses of both kinds begin to recur again in the same order that they occur in the next preceding period. This is called the saros, and it was known to the Chaldeans 2600 years ago.

6. The Planets. We have several times mentioned the fact that, beside the earth, there are seven other principal planets revolving round the sun, in the same direction as the earth, but at various distances. We shall consider each of these in the order of its distance from the sun.

But first it is desirable to explain briefly certain so-called "laws" which govern the motions of all the planets. These are known as Kepler's laws of planetary motion, and are three in number. The demonstration of their truth would carry us beyond the scope of this book, and consequently we shall merely state them as they are recognised by astronomers.

1st Law: The orbit of every planet is an ellipse, having the sun situated in one of the foci.

2d Law: The radius vector of a planet describes equal areas in equal times. By the radius vector is meant the straight line joining the planet to the sun, and the law declares that as the planet moves round the sun, the area of space swept over by this line in any given time, say one day, is equal to the area which it will sweep over in any other equal length of time. If the orbit were a circle it is evident at a glance that the law must be true, because then the sun would be situated in the centre of the circle, the length of the radius vector, no matter where the planet might be in the orbit, would never vary, and the area swept over by it in one day would be equal to the area swept over in any other day, because all these areas would be precisely similar and equal triangles. But Kepler discovered that the same thing is true when the orbit is an ellipse, and when, in consequence of the eccentricity of the orbit, the planet is sometimes farther from the sun than at other times. As the triangular area swept over in a given time increases in length with the planet's recession from the sun, it diminishes in breadth just enough to make up the difference which would otherwise exist between the different areas. This law grows out of the fact that the force of gravitation varies inversely with the square of the distance.

3d Law: The squares of the periods (*i. e.* times of revolution in their orbits) of the different planets are proportional to the cubes of their mean (average) distances from the sun. The meaning of this will be best explained by an example. Suppose one planet, whose distance we know, has a period only one-eighth as long as that of another planet, whose distance we do not know. Then Kepler's third law enables us to calculate the distance of the second planet. Call the period of the first planet 1, and that of the second 8, and also call the distance of the first 1, since all we really need to know is the *relative* distance of the second, from which its distance in miles is readily deduced by comparison with the distance of the first. Then, by the law, 12 : 82 : 13 : x3 ("x" representing the unknown quantity). Now, this is simply a problem in proportion where the product of the means is equal to the product of the extremes. But 12 = 1, and also 13= 1; therefore x3 = 82, and x = (82) (the cube root of the square of 8), which is 4. Thus we see that the distance of the second planet must be four times that of the first.

This third law of Kepler is applied to ascertain the distances of newly discovered planets, whose periods are easily ascertained by simple observation. If we know the distance of any one planet by measurement, we can calculate the distances of all the others after observing their periods. The law also works conversely, *i. e.* from the distances the periods can be calculated.

We now take up the various planets singly. The nearest to the sun, as far as known, is Mercury, its average distance being only 36,000,000 miles. But its orbit is so eccentric that the distance varies from 28,500,000 miles at perihelion to 43,500,000 at aphelion. In consequence its speed in its orbit is very variable, and likewise the amount of heat and light received by it from the sun. On the average it gets more than 6½ times as much solar light and heat as the earth gets. But at perihelion it gets 2½ times as much as at aphelion, and the time which it occupies in passing from perihelion to aphelion is only six weeks, its entire year being equal to 88 of our days. Being situated so much nearer the sun than the earth is, Mercury is never visible to us except in the morning or the evening sky, and then not very far from the sun. Its diameter is about 3000 miles, but its mass is not certainly known from lack of knowledge of its mean density. This lack of knowledge is due to the fact that Mercury has no satellite. When a planet has a satellite it is easy to calculate its density from its measured diameter combined with the orbital speed of its satellite. Certain considerations have led some to believe that the mean density of Mercury may be very great, perhaps as great as that of lead, or of the metal mercury itself. Not knowing the mass, we cannot say exactly what the weight of bodies on Mercury is. We are also virtually ignorant of the condition of the surface of this planet, the telescope revealing very little detail, but it is generally thought that it bears a considerable resemblance to the surface of the moon. There is another way in which Mercury is remarkably like the moon. The latter, as we have seen, always keeps the same side turned toward the earth, which is the same thing as saying that it turns once on its axis, while going once around the earth. So Mercury keeps always the same side toward the sun, making one rotation on its axis in the course of one revolution in its orbit. Consequently, one side of Mercury is continually in the sunlight, while the opposite side is continually buried in night. There must, however, be regions along the border between these two sides, where the sun does rise and set once in the course of one of Mercury's years. This arises from the eccentricity of the orbit, and the consequent variations in the orbital velocity of the planet, which cause now a little of one edge and now a little of the other edge of the dark hemisphere to come within the line of sunlight. (The same thing occurs with the moon, though to a less degree owing to the smaller eccentricity of the moon's orbit, which, however, is sufficient to enable us to see at one time a short distance round one side of the moon and at another time a short distance round the opposite side.) This phenomenon is known as libration. Mercury apparently possesses an atmosphere, but we know nothing certain concerning its density.

The next planet, in the order of distance from the sun, is Venus, whose average distance is 67,200,000 miles. The orbit of Venus is remarkable for its small eccentricity, so that the difference between its greatest and least distances from the sun is less than a million miles. The period, or year, of Venus is 225 of our days. Owing to her situation closer to the sun, she gets nearly twice as much light and heat as the earth gets. In size Venus is remarkably like the earth, her diameter being 7713 miles, which differs by only 205 miles from the mean diameter of the earth. Her axis is nearly perpendicular to the plane of her orbit. Her globe is a more perfect sphere than that of the earth, being very little flattened at the poles or swollen at the equator. Although Venus, like Mercury, has no satellite, her mean density has been calculated by other means, and is found to be 0.89 that of the earth. From this, in connection with her measured diameter, it is easy to deduce her mass, and the force of gravity on her surface. The latter comes out at about 0.85 that of the earth, *i. e.* a body weighing 100 pounds on the earth would weigh 85 pounds if removed to Venus. She possesses an atmosphere denser and more extensive than would theoretically have been expected—indicating, perhaps, a difference of constitution. Her atmosphere has been estimated to be twice as dense as ours, a great advantage, it may be remarked, from the point of view of aëronautics. But this dense and abundant atmosphere renders Venus a very difficult object for the telescope on account of the brilliance of its reflection. In consequence, we know but little of the surface of the planet.

One important result of this is that the question remains undecided whether Venus rotates on her axis at a rate closely corresponding with that of the earth, as some observers think, or whether, as others think, she, like Mercury, turns only once on her axis in going once round the sun. The importance of the question in its bearing on the habitability of Venus is apparent, for if she keeps one face always sunward, then on one side there is perpetual day and on the other perpetual night. On the other hand, if she has days and nights approximately equal in length to those of the earth, it may well be thought that she is habitable by beings not altogether unlike ourselves, because the force of gravity on her surface is not much less than on the earth, and her dense atmosphere, filled with clouds, might tend to shield her inhabitants from the effects of the greater amount of heat poured upon her by the sun. As her orbit is inside that of the earth, Venus, like Mercury, is only visible either in the evening or in the morning sky, but owing to her greater actual distance from the sun, her apparent distance from it in the sky is greater than that of Mercury.

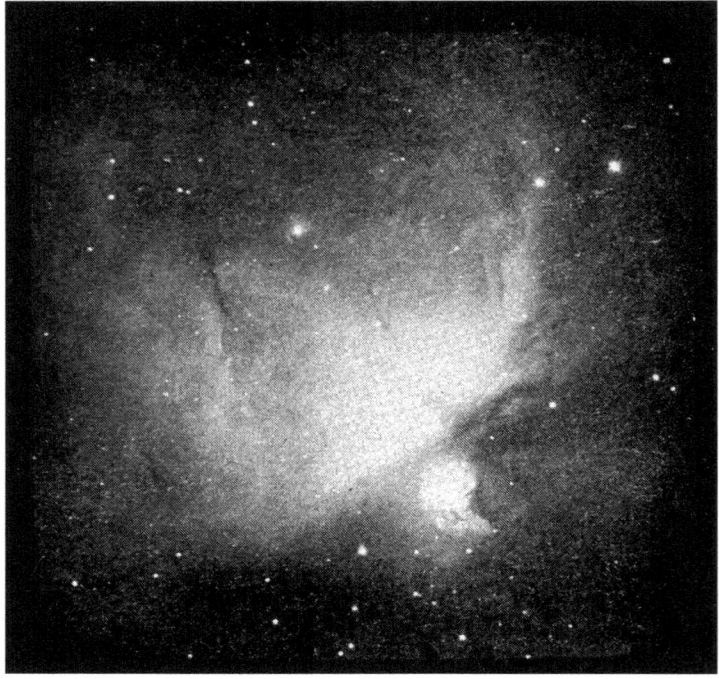

The Great Nebula in Orion

Photographed at the Lick Observatory by J. E. Keeler, with the Crossley reflector. Exposure one hour.

Both of these planets, in consequence of passing alternately between the sun and the earth and round the opposite side of the sun, present phases resembling those of the moon. The reader can explain these to himself by means of the experiment, before mentioned, with a billiard ball and a lamp. In this case let the observer remain seated in his chair while another person carries the ball round the lamp in such a manner that it shall alternately pass between the lamp and the observer and round the other side of the lamp. When Venus comes nearly in line between the earth and the sun, she becomes an exceedingly brilliant object in either the evening or the morning sky, although at such times we see, in the form of a crescent, only a part of that half of her surface which is illuminated. Her increase of brightness at such times is due to her greater nearness to the earth. When between the earth and the sun she may be only about 26,000,000 miles away, while when she is on the other side of the sun she may be over 160,000,000 miles away. Both

Venus and Mercury when passing exactly between the sun and the earth are seen, in the form of small black circles, moving slowly across the sun's disk. These occurrences are called transits, and in the case of Venus have been before referred to. They are more frequent with Mercury than with Venus, but Mercury's transits are not utilisable for parallax observations. The latest transit of Venus occurred in 1882, and there will not be another until 2004. The latest transit of Mercury occurred in 1907, and there will be another in 1914.

The earth is the third planet in order of distance, and then comes Mars, whose average distance from the sun is 141,500,000 miles. The orbit of Mars is so eccentric that the distance varies between 148,000,000 and 135,000,000 miles. Its period or year is about 687 of our days. In consequence of its distance, Mars gets, on the average, a little less than half as much light and heat as the earth gets. When it is on the same side of the sun with the earth, and nearly in line with them, it is said to be in opposition. At such times it is manifestly as near the earth as it can come, and thus an opposition of Mars offers a good opportunity for the telescopic study of its surface. These oppositions occur once in about 780 days, but they are not all of equal importance, because the distance between the two planets is not the same at different oppositions. The cause of the difference of distance is the eccentricity of the orbit. If an opposition occurs when Mars is in aphelion its distance from the earth will be about 61,000,000 miles, but if the opposition occurs when Mars is in perihelion the distance will be only about 35,000,000 miles. The average distance at an opposition is about 48,500,000 miles. The most favourable oppositions always occur in August or September, and are repeated at an interval of from fifteen to seventeen years. But at some of the intervening oppositions the distance of the planet is not too great to afford good views of its surface. The diameter of Mars is about 4330 miles, with a similar polar flattening to that of the earth. Its density is 0.71 that of the earth, and the force of gravity on its surface 0.38. A body weighing 100 pounds on the earth would weigh 38 pounds on Mars. The evidence in regard to its atmosphere is conflicting, but the probability is that it has an atmosphere not denser than that existing on our highest mountain peaks. Opinions concerning the existence of water vapour on Mars are also conflicting. One fact tending to show that its atmosphere must be very rare and cloudless is that its surface features are very plainly discernible with telescopes.

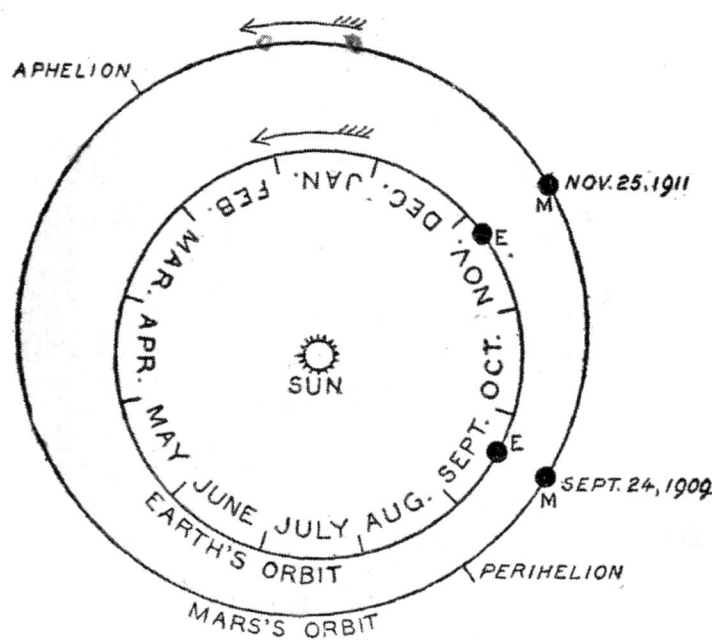

Fig. 16. Orbits of Mars and the Earth.

Inspection shows at once why the oppositions of Mars which occur in August and September are the most favourable because Mars being then near the perihelion point of its elongated orbit is comparatively near the earth, while oppositions which occur in February and March are very unfavourable because then Mars is near the aphelion point of its orbit, and its distance from the earth is much greater. The oppositions occur along the more favourable part of the orbit about two years and two months apart. Thus the figure shows that the opposition of September 24, 1909 was followed by one on November 25, 1911.

About each pole, as it happens to be turned earthward, is to be seen a round white patch (supposed to be snow), and this gradually disappears as the summer advances in that hemisphere of the planet—for Mars has seasons very closely resembling our seasons, except that they are about twice as long. The inclination of the axis of Mars to the plane of its orbit is about 24° 50 , which is not very different from the inclination of the earth's axis. Moreover, Mars rotates in a period of 24 hours, 37 min., 22 sec., so that the length of day and night upon its surface is very nearly the same as upon the earth. The surface of the planet is marked by broad irregular areas of contrasting colour, or tone, some of them being of a slightly reddish, or yellowish, hue, and others of a neutral dusky tint. The general resemblance to a globe of the earth, with differently shaped seas and oceans, is striking.

On account of the many likenesses between Mars and the earth, some astronomers are disposed to think that Mars may be a habitable planet. The terms "seas" and "continents" were formerly applied to the contrasted areas just spoken of, but now it is believed that there are no large bodies of water on Mars. Crossing the light, or reddish-coloured, areas there are sometimes seen great numbers of intersecting lines, very narrow and faint, which have received the name of "canals." Some speculative minds find in these ground for believing that they are of artificial origin, and a theory has been built up, according to which the so-called canals are "irrigated bands," the result of the labours of the inhabitants. The argument of the advocates of this theory is put about as follows: Mars is evidently a nearly dried-up planet, and most of the water left upon it is periodically locked up in the polar snows. As these snows melt away in the summer time, now in one hemisphere and now in the other, the water thus formed is conducted off toward the tropical and equatorial zones by innumerable canals, too small to be seen from the earth. The lands irrigated by these canals are narrow strips, whose situation is determined by local circumstances, and which cross one another in all directions. Within these bands, which enlarge into rounded "oases" where many of them intersect, vegetation pushes, and its colour causes them to appear as dark lines and patches on the surface of the planet. The fact that the lines make their appearance gradually, after the polar caps begin to disappear, is regarded as strongly corroborative of the theory. In answer to the objection that works so extensive as this theory of irrigation calls for would be practically impossible, it is replied that the relatively small force of gravity on Mars not only immensely diminishes the weight of all bodies there, but also renders it possible for animal forms to attain a greater size, with corresponding increase of muscular power. It is likewise argued that Mars may have been longer inhabited than the earth, and that its inhabitants may consequently have developed a more complete mastery over the powers of nature than we as yet possess. Many astronomers reject these speculations, and even aver that the lines called "canals" (and it must be admitted that many powerful telescopes show few or none of them) have no real existence, what is seen, or imagined to be seen, being due to some peculiarity of the soil, rocks, or atmosphere.

Mars has two small satellites, revolving round it with great speed at close quarters. The more distant satellite, Deimos, is 14,600 miles from the centre of Mars and goes round it in 30 hours, 18 min. The nearer one, Phobos, is only 5800 miles from the planet's centre, and its period of revolution is only 7 hours, 39 min., so that it makes more than three circuits while the planet is rotating once on its axis. Both of the satellites are minute in size, probably under ten miles in diameter.

Beyond Mars, at an average distance of about 246,000,000 miles from the sun, is a system of little planets called asteroids. More than 600 are now known, and new ones are discovered every year, principally by means of photography. Only four of these bodies are of any considerable size, and they were, naturally, the first to be discovered. They are Ceres, diameter 477 miles; Pallas, 304 miles; Vesta, 239 miles; and Juno, 120 miles. Many of the others have a diameter of only about ten, or even, perhaps, as little as five, miles. Their orbits are more eccentric than those of any of the large planets, and one of them, Eros, has a mean distance of 135,000,000 miles, and a least distance of only 105,000,000, so that it is nearer to the sun than Mars is. Eros may, under favourable circumstances, approach within 14,000,000 miles of the earth. This fact, as already mentioned, has been taken advantage of for measuring its distance from the earth, from which the distance of the sun may be calculated with increased accuracy. Eros and some others of the asteroids seem to be of an irregular or fragmentary form, and this has been used to support a theory, which is not, however, generally accepted, that the asteroids are the result of an explosion, by which a larger planet was blown to pieces.

Sixth in order of distance from the sun (counting the asteroids as representing a single body) is the greatest of all the planets, Jupiter. His average distance from the sun is 483,000,000 miles, but the eccentricity of his orbit causes him to approach within 472,500,000 miles at perihelion, and to recede to 493,500,000 miles at aphelion. When in opposition, Jupiter's mean distance from the earth is 390,000,000 miles. This gigantic planet has a mean diameter of 87,380 miles, but is so flattened at the poles and bulged round the equator that the polar diameter is only 84,570 miles, while the equatorial diameter is 90,190 miles, a difference of 5680 miles. This peculiar form is doubtless due to the planet's swift rotation. The axis, like that of Venus, is nearly perpendicular to the plane of the orbit. He makes a complete turn on his axis in a mean period of 9 hours, 55 minutes. The reason for saying "a mean period" will appear in a moment. Jupiter's year is equal to 11.86 of our years, but it comes into opposition to the sun, as seen from the earth, once in every 399 days.

The volume of Jupiter is about 1300 times that of the earth, *i.e.* it would take 1300 earths rolled into one to equal Jupiter in size. But its mean density is slightly less than one quarter of the earth's, so that its mass is only 316 times greater than the earth's. The force of gravity on its surface is 2.64 times the earth's. A body weighing 100 pounds on the earth would weigh 264 pounds on Jupiter. It will be observed that Jupiter's mean density is very nearly the same as that of the sun, and we conclude that it cannot be a solid, rigid globe like the earth. This conclusion is made certain by the fact that its period of rotation on its axis is variable, another resemblance to the sun. The equatorial parts go round in a shorter period than parts situated some distance north or south of the equator. It may be supposed that there is a solid nucleus within, but if so, no direct evidence of its existence has been found.

Nevertheless, although Jupiter appears to be in a cloud-like state, it does not shine with light of its own, so that its temperature, while no doubt higher than that of the earth, cannot approach anywhere near that of the sun. We do not know of what materials Jupiter is composed, for spectroscopic analysis applies especially to bodies which shine with their own light. When they shine only by reflected light received from the sun, their spectra resemble the regular solar spectrum, except for the presence of faint bands due to absorption in the planet's atmosphere. It may be that there are no elements of great atomic density, such as iron or lead, in the globe of Jupiter. Yet in the course of long ages the planet may become smaller and more condensed, in consequence of the escape of its internal heat. In this way Jupiter may be regarded as representing an intermediate stage of evolution between an altogether vaporous and very hot body like the sun, and a cool and solid one like the earth.

Jupiter presents a magnificent appearance in a good telescope. Its oblong disk is seen crossed in an east and west direction, and parallel to its equator, by broad, vari-coloured bands, called belts. These frequently change in form and, to some extent, in situation, as well as in number. But there are always at least two wide belts, one on each side of the equator. In 1878 a very remarkable feature was noticed just south of the principal south belt of Jupiter, which has become celebrated under the name of the Great Red Spot. In a few years after its discovery its colour faded, but it still remains visible, with varying degrees of distinctness, as an oblong marking, about 30,000 miles long and 7000 miles broad. The outer border of the great south belt bends away from the spot, as if some force of repulsion acted between them, or as if the spot were an elevation round which the clouds of the belt flowed like a river round a projecting headland. The nature of this curious spot is unknown. Other smaller spots, sometimes white, sometimes dusky, occasionally make their appearance, but they do not exhibit the durability of the Great Red Spot.

Jupiter has eight satellites, four of which, known since the time of Galileo, are conspicuous objects in the smallest telescope. All but one of these four are larger than our moon, while the other four are extremely insignificant in size. The four principal satellites are designated by Roman numerals, I, II, III, IV, arranged in the order of distance from the planet. They also have names which are seldom used. Satellite I (Io) has a diameter of 2452 miles, and revolves in a period of 1 day, 18 hours, 27 min., 35.5 sec., at a mean distance of 261,000 miles; II (Europa) is 2045 miles in diameter, and revolves in 3 days, 13 hours, 13 min., 42.1 sec., at a mean distance of 415,000 miles; III (Ganymede) has a diameter of 3558 miles, a period of seven days, 3 hours, 42 min., 33.4 sec., and a mean distance of 664,000 miles; IV (Callisto) is 3345 miles in diameter, has a period of 16 days, 16 hours, 32 min., 11.2 sec., and a mean distance of 1,167,000 miles. The object of giving the periods with extreme accuracy will appear when we speak of the use made of observations of Jupiter's satellites. The first of the four small satellites, discovered by Barnard in 1892, is probably less than 100 miles in diameter, and has a mean distance of 112,500 miles, and a period of only 11 hours, 57 min., 22.6 sec. The other small satellites are much more distant than any of the large ones, the latest to be discovered, the eighth, being situated at a mean distance of about 15,000,000 miles, but travelling in an orbit so eccentric that the distance ranges between 10,000,000 and 20,000,000 miles. The period is about two and a fifth years. But the most remarkable fact is that this satellite revolves round Jupiter from east to west, a direction contrary to that pursued by all the others, and contrary to the direction which is almost universal among the rotating and revolving bodies of the solar system.

The large satellites are very interesting objects for the telescope. When they come between the sun and Jupiter their round black shadows can be plainly seen moving across his disk, and when they pass round into his shadow they are suddenly eclipsed, emerging after a time out of the other side of the shadow. These phenomena are known as transits and eclipses, and their times of occurrence are carefully predicted in the *American Ephemerisand Nautical Almanac*, published at Washington for the benefit of astronomers and navigators, because these eclipses can be employed in comparing local time with standard meridian time. They were formerly utilised to determine the velocity of light, in this way:

As the earth goes round its orbit inside that of Jupiter the latter is seen in opposition to the sun at intervals of 399 days. When it is thus seen the earth must be between the sun and Jupiter, and the distance between the two planets is the least possible. But when the earth has passed round to the other side of the sun from Jupiter this distance becomes the greatest possible. The increase of distance between the two planets, as the earth goes from the nearest to the farthest side of its orbit, is about 186,000,000 miles. Now it was noticed by the Danish astronomer, Roemer, that as the earth moved farther and farther from Jupiter the times of occurrence of the eclipses kept getting later and later, until when the earth arrived at its greatest distance the eclipses were about 16 minutes behind time. He correctly inferred that the retardation of the time was due to the increase of the distance, and that the 16 minutes by which the eclipses were behindhand when the distance was greatest represented the time taken by light to cross the 186,000,000 miles of space by which the earth had increased its distance from Jupiter. In other words, light must travel 186,000,000 miles in about sixteen minutes, from which it was easy to calculate its speed per second—which we now know to be 186,330 miles. Our knowledge of the velocity of light furnishes one of the means of calculating the distance of the sun.

We come next to the beautiful planet Saturn, whose mean distance from the sun is 886,000,000 miles. The distance varies between 911,000,000 and 861,000,000 miles. Saturn's year is equal to 29.46 of our years. It comes into opposition every 378 days. The most surprising feature of Saturn is the system of immense rings surrounding it above the equator. The globe of the planet is 76,470 miles in equatorial diameter, and 69,780 miles in polar diameter, a difference of 6690 miles, so that Saturn is even more compressed at the poles

and swollen at the equator than Jupiter. The axis of rotation is inclined 27°
from a perpendicular. The rings are three in number, very thin in proportion
to their vast size, and placed one within another in the same plane. The outer
diameter of the outer ring, called Ring A, is about 168,000 miles. Its breadth
is about 10,000 miles. Then comes a gap, about 1600 miles across, separating
it from Ring B, the brightest of the set. This is about 16,500 miles broad,
and at its inner edge it gradually fades out, blending with Ring C, which
is called the crape, or gauze, ring, because it has a dusky appearance, and
is so translucent that the globe of the planet can be seen through it. This
ring is about 10,000 miles broad, and its inner edge comes within a distance of
between 9000 and 10,000 miles of the surface of the planet. Ring A apparently
has a very narrow gap running round at about a third of its breadth from the
outer edge. This, known as Encke's Division, is not equally plain at all times.
Occasionally observers report the temporary appearance of other thin gaps.

The mean density of Saturn is less than that of any other planet, being but
0.13 that of the earth, or 0.72 that of water. It follows that this great planet
would float in water. The weight of bodies at its surface would be a little
less than three-quarters of their weight on the surface of the earth. The globe
of Saturn, like that of Jupiter, is marked by belts parallel with the equator,
but they are less definite in outline and less conspicuous than the belts of
Jupiter. The equatorial zone often shows a beautiful pale salmon tint, while
the regions round the poles are faintly bluish. Light spots are occasionally
seen upon the planet, and it appears to rotate more rapidly at the equator
than in the higher latitudes. There seems to be every reason to think that
Saturn, also, is of a vaporous constitution, although it may have a relatively
condensed nucleus.

But while the globe of the planet appears to be vaporous, the same is
not true of the rings. We have already mentioned the fact that they are
exceedingly thin in proportion to their great size and width. The thickness
has not been determined with exactness, but it probably does not exceed,
on the average, one hundred miles. There appear to be portions of the
rings which are thicker than the average, as if the matter of which they
are composed were heaped up there. This matter evidently consists of an
innumerable multitude of small bodies. In other words, the rings are composed
of swarms of what may be called meteors. That their composition must be
of this nature, although the telescope does not reveal it, has been proved in
two ways: first, by mathematical calculation, which shows that if the rings
were all of a piece, whether solid or liquid, they would be destroyed by the
contending forces of attraction to which they are subject; and, second, by
spectroscopic observation, which proves, in a way that will be shown when we
come to deal with the stars, that the rings rotate with velocities proportional
to the distances of their various parts from the centre of the planet. Hence it
is inferred that they must consist of a vast number of small bodies or particles.

Saturn has ten satellites, all revolving outside the rings. The names of nine of these in the order of increasing distance are: Mimas, distance 117,000 miles; Enceladus, distance 157,000 miles; Tethys, distance 186,000 miles; Dione, distance 238,000 miles; Rhea, distance 332,000 miles; Titan, distance 771,000 miles; Hyperion, distance 934,000 miles; Japetus, distance 2,225,000 miles; and Phœbe, distance 8,000,000 miles. The last, like the eighth satellite of Jupiter, revolves in a retrograde direction. Only Titan and Japetus are conspicuous objects. The period of Mimas is only about 22½ hours; that of Titan is 15 days, 22 hours, 41 min., and that of Japetus about 79 days, 8 hours. Barnard's measurements indicate for Titan a diameter of 2720 miles. Japetus is probably about two-thirds as great in diameter as Titan.

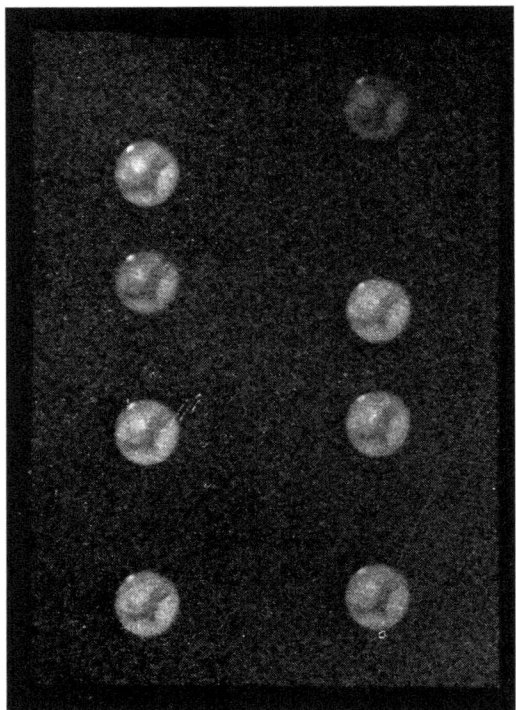

Photographs of Mars

Made at the Yerkes Observatory by E. E. Barnard, with the forty-inch refractor, September 28, 1909.

Beyond Saturn, in the order named, are Uranus and Neptune. The mean distance of the former from the sun is 1,782,000,000 miles, and that of the latter 2,791,500,000 miles. The orbit of Uranus is more eccentric than that of Neptune. The diameter of Uranus is about 32,000 miles and that of Neptune about 35,000 miles. The year of Uranus is equal to 84 of our years, and that of Neptune to 164.78. These planets are so remote, and so poorly illuminated by the sun, that the telescope reveals very little detail on their surfaces. Their density is somewhat less than that of Jupiter. Uranus has four satellites, Ariel, Umbriel, Titania, and Oberon, situated at the respective distances of 120,000, 167,000, 273,000, and 365,000 miles. Neptune has one, nameless, satellite, at a distance of 225,000 miles.

The most remarkable thing about these two planets is that their axes of rotation, as compared with those of all the other planets, are tipped over into a different plane, so that they rotate in a retrograde or backward direction, and their satellites, in like manner, revolve from east to west. The axis of Uranus is not far from upright to the plane of the ecliptic, so that the motion of its satellites carries them alternately far northward and far southward of that plane, but the axis of Neptune is tipped so far over that the retrograde, or east to west, motion is very pronounced. Neptune is celebrated for having been discovered by means of mathematical calculations, based on its disturbing attraction on Uranus. These calculations showed where it ought to be at a certain time, and when telescopes were pointed at the indicated spot the planet was found. Similar disturbances of the motions of Neptune lead some astronomers to think that there is another, yet undiscovered, planet still more distant.

Comets. Comets are the most extraordinary in appearance of all celestial objects visible to the naked eye. Great comets have been regarded with terror and superstitious dread in all ages of the world, wherever ignorance of their nature has prevailed. They have been taken for prognosticators of wars, famines, plagues, the death of rulers, the outbreak of revolutions, and the subversion of empires. One reason for this, aside from their strange and menacing appearance, is, no doubt, the rarity of very great and conspicuous comets. It was not until Newton had demonstrated the law of gravitation that the fact began to be recognised that comets are controlled in their motions by the sun. We now know that they travel in orbits, frequently, and perhaps always, elliptical, having the sun in one of the foci. Comets are habitually divided into two classes: first, periodical comets, meaning those which have been observed at more than one return to perihelion; and, second, non-periodical comets, meaning those which have been seen but once, but which, nevertheless, may return to perihelion in a period so long that a second return has not been observed. A better division is into comets of short period, and comets of long, or unknown, periods.

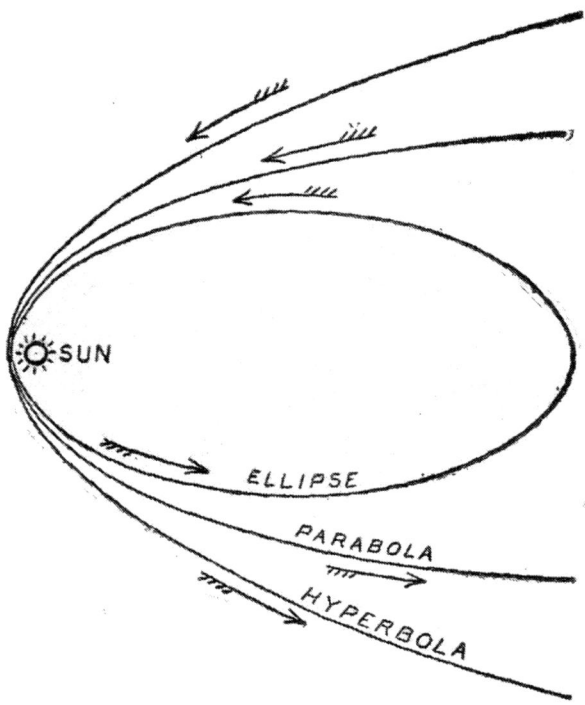

Fig. 17. Ellipse, Parabola, and Hyperbola.

The figure shows graphically why it is so difficult to tell exactly the form of a comet's orbit. The three kinds of curves are nearly of the same form near the focus (the Sun), and it is only in that part of its orbit that the comet can be seen. Moreover a comet is, at best, a misty and indefinite object, which renders it so much the more difficult to obtain good observations of its precise position and movement.

Still, many astronomers are disposed to think that the majority of comets do not travel in elliptical but in parabolic, and a few in hyperbolic, orbits. This calls for a few words of explanation. Ellipses, parabolas, and hyperbolas are all conic-section curves, but the ellipse alone returns into itself, or forms a closed circuit. In each case the sun is situated at the focus where the perihelion, or nearest approach, of the comet occurs, but only comets travelling in elliptical orbits return again after having once been seen. A comet moving in a parabola would go back into the depths of space nearly in the direction from which it had come, and would never be seen again; and if it moved in a hyperbola it would go off toward another quarter of the celestial sphere, and likewise would never return. Now it is true that the forms calculated for the orbits of the majority of comets that have been observed appear to be parabolic (a very few seem to be hyperbolic), and if this is the fact such comets cannot be permanent members of the solar system, but must enter it from far-off regions of space, and having visited the sun must return to such regions without any tendency to come back again. In that case they may pay similar visits to other suns.

But it is quite possible that what appear to be parabolic orbits may, in reality, be ellipses of very great eccentricity. The difficulty in determining the precise shape of a comet's orbit arises from the fact that all three of the curves just mentioned closely approximate to one another in the neighbourhood of their common focus, the sun, and it is only in that part of their orbits that comets are visible. The whole question is yet in abeyance, but, as we have said, it seems likely that all comets really move in elliptical orbits, and consequently never get entirely beyond the control of the sun's attraction. But in all cases the orbits of comets are much more eccentric than those of the large planets. The famous comet of Halley, for instance, which has the longest period of any of the known periodical class, about seventy-five years, is 3,293,000,000 miles from the sun when in aphelion, and only 54,770,000 miles when in perihelion.

Comets, when near the sun, are greatly affected by the disturbing attraction of large planets, and especially of the most massive of them all, Jupiter. The effect of this disturbance is to change the form of their orbits, with the not infrequent result that the latter are altered from apparent parabolas into unquestionable ellipses, and thus the comets concerned are said to be "captured," or made prisoners to the sun, by the influence of the disturbing planet. About twenty small comets are known as "Jupiter's Comet Family," because they appear to have been "captured" in this way by him. A few others are believed to have been similarly captured by Saturn, Uranus, and Neptune.

The orbits of comets differ from those of the planets in other ways beside their greater eccentricity. The planets all move round the sun from west to east, but comets move in both directions. The orbits of the planets, with the exception of some of the asteroids, all lie near one common plane, but those of comets are inclined at all angles to this plane, some of them coming down from the north side of the ecliptic and others up from the south side.

A comet consists of two distinct portions: first, the head, or nucleus; and, second, the tail. The latter only makes its appearance when the comet is drawing near the sun, and, as a whole, it is always directed away from the sun, but usually more or less curved backward along the comet's course, as if the head tended to run away from it. The appearance of a comet's tail at once suggests that it is produced by some repulsive force emanating from the sun. Recently there has been a tendency to explain this on the principle of what is known as the pressure of light. This demands a brief explanation. Light is believed to be a disturbance of the universal ether in the form of waves which proceed from the luminous body. These waves possess a certain mechanical energy tending to drive away bodies upon which they impinge. The energy is relatively slight, and in ordinary circumstances produces no perceptible effect, but when the body acted upon by the light is extremely small the pressure may become so great relatively to gravitation as to prevail over the latter. To illustrate this, let us recall two facts—first, that gravitation acts upon the mass, *i.e.* all the particles of a body throughout its entire volume; and, second, that pressure acts only upon the exterior surface. Consequently gravitation is proportional to the volume, while the pressure of light is proportional to the surface of the body acted upon. Now the mass, or volume, of any body varies as the cube of its diameter, and the surface only as the square. If, then, we have two bodies, one of which has twice the diameter of the other, the mass of the second will be eight times less than that of the first, but the surface will be only four times less. If the second has only one-third the diameter of the first, then its mass will be twenty-seven times less, but its surface only nine times less. Thus we see that as we diminish the size of the body, the mass falls off more rapidly than the area of the surface, and consequently the pressure gains relatively to the gravitation. Experiment has corroborated the conclusions of mathematics on this subject, and has shown that when a particle of matter is only about one one-hundred-thousandth of an inch in diameter the pressure of light upon it becomes greater than the force of gravitation, and such a particle, situated in open space, would be driven away from the sun by the light waves. This critical size would vary with the density of the matter composing the particle, but what we have said will serve to convey an idea of its minuteness.

Now in applying this to comets' tails it is only necessary to remark that they are composed of either gaseous or dusty particles, or both, rising from the nucleus, probably under the influence of the heat or the electrical action of the sun, and these particles, being below the critical size, are driven away from the sun, and appear in the form of a tail following the comet. It may be added that the same principle has been evoked to explain the corona of the sun, which may be composed of clouds of gas or dust kept in suspension by the pressure of light.

The nuclei of comets contain nearly their whole mass. The actual mass of no comet is known, but it can in no case be very great. Moreover, it is probable that the nucleus of a comet does not consist of a single body, either solid or liquid, but is composed of a large number of separate small bodies, like a flock of meteors, crowded together and constantly impinging upon one another. As the comet approaches the sun the nucleus becomes violently agitated, and then the tail begins to make its appearance.

The possibility exists of an encounter between the earth and the head of a comet, but no such occurrence is known. Two or three times, however, the earth is believed to have gone through the tail of a comet, the last time in 1910, when Halley's comet passed between the earth and the sun, but no certain effects have been observed from such encounters. The spectroscope shows that comets contain various hydrocarbons, sodium, nitrogen, magnesium, and possibly iron, but we know, as yet, very little about their composition. The presence of cyanogen gas was reported in Halley's comet at its last appearance.

We are still more ignorant of the origin of comets. We do know, however, that they tend to go to pieces, especially those which approach very close to the sun. The great comet of 1882, which almost grazed the sun, was afterward seen retreating into space scattered into several parts, each provided with a tail. In at least one case, several comets have been found travelling in the same track, an indication that one large original comet has been separated into three or four smaller ones. This appears to be true of the comets of 1843, 1880, and 1882,—and perhaps the comet of 1576 should be added. But the most remarkable case of disruption is that of Biela's Comet, which first divided into two parts in 1846 and then apparently became scattered into a swarm of meteors which was encountered by the earth in 1872, when it passed near the old track of the comet. This leads us to our next subject.

8.   Meteors.   Everybody must, at some time, have beheld the phenomenon known as a falling, or shooting, star. A few of these objects can be seen darting across the sky on almost any clear night in the course of an hour or two of watching. Sometimes they appear more numerously, and at intervals they are seen in "showers." They are called meteors, and it is believed that they are minute solid bodies, perhaps averaging but a small fraction of an ounce in weight, which plunge into the atmosphere with velocities varying from twenty to thirty or more miles per second, and are set afire and consumed by the heat of friction developed by their rush through the air. Anybody who has seen a bullet melted by the heat suddenly developed when it strikes a steel target has had a graphic illustration of the transformation of motion into heat. But if we could make the bullet move fast enough it would *melt in the air*, the heat being developed by the constant friction.

The connection of meteors with comets is very interesting. In the year 1833, a magnificent and imposing display of meteors, which, for hours, on the night between the 13th and 14th of November, filled the sky with fire-balls and flaming streaks, astonished all beholders and filled many with terror. It was found that these meteors travelled in an orbit intersecting that of the earth at the point where the latter arrived in the middle of November, and also that they had a period of revolution about the sun of 33¼ years, and were so far scattered along their orbit that they required nearly three years to pass the point of intersection with the orbit of the earth. Thus it was concluded that for three years in succession, in mid-November, there should be a display of the meteors plunging into the earth's atmosphere. But only in the year when the thickest part of the swarm was encountered by the earth would the display be very imposing. Upon this it was predicted that there would be a recurrence of the phenomenon of 1833 in the year 1866. It happened as predicted, except that the number of meteors was not quite so great as before. In the meantime, it had been discovered that these meteors followed in the track of a comet known as Temple's Comet, and also that certain other meteors, which appear every year in considerable numbers about the 10th of August, followed the track of another comet called Tuttle's Comet. Then in

1872 came the display, mentioned in the last section, of meteors which were evidently the debris of the vanished comet of Biela. The inference from so many similar cases was irresistible that the meteors must be fragments of destroyed or partially destroyed comets. Several other cases of identity of orbits between meteors and comets have been discovered.

It has been said that the August meteors appear every year. The explanation of this is that they have, in the course of many ages, been scattered around the whole circuit of their orbit, so that each year, about the 10th of August, when the earth crosses their track, some of the meteors are encountered. They are like an endless railroad train travelling upon a circular track. The November meteors also appear, in small numbers, every year, a fact indicating that some of them, too, have been scattered all around their orbit, although the great mass of them is still concentrated in an elongated swarm, and a notable display can only occur when this swarm is at the crossing simultaneously with the earth. These meteors were eagerly awaited in 1899, when it was hoped that the splendid displays of 1833 and 1866 might be repeated, but, unfortunately, in the meantime the planets Jupiter and Saturn, by their disturbing attractions, had so altered the position of the path of the meteors in space that the principal swarm missed the connection. There are many other periodical meteor showers, generally less brilliant than those already mentioned, and some astronomers think that all of them had their origin from comets.

It is not known that any meteor from any of these swarms has ever reached the surface of the earth. The meteors appear to be so small that they are entirely burnt up before they can get through the atmosphere, which thus acts as a shield against these little missiles from outer space. But there is another class of meteoric bodies, variously known as meteorites, aërolites, uranoliths, or bolides, which consists of larger masses, and these sometimes fall upon the earth, after a fiery passage through the air. Specimens of them may be seen in many museums. They are divided into two principal classes, according to their composition: first, stony meteorites, which are by far the most numerous; and, second, iron meteorites, which consist of almost pure iron, generally alloyed with a little nickel. The stony meteorites, which usually contain some compound of iron, consist of a great variety of substances, including between twenty and thirty different chemical elements. Although they resemble in many ways minerals of volcanic origin on the earth, they also possess certain characteristics by which they can be recognised even when they have not been seen to fall.

When a meteorite passes through the air it makes a brilliant display of light, and frequently bursts asunder, with a tremendous noise, scattering its fragments about. The largest fragment of a meteorite actually seen to fall, weighs about a quarter of a ton. Upon striking the ground the meteorite sometimes penetrates to a depth of several feet, and some have been picked up which were yet hot on the surface, although very cold within. It is not known that meteorites have any connection with comets, and their origin can only be conjectured. Among the various suggestions that have been made the following may be mentioned: (1) that they have been shot out of the sun—particularly the iron meteorites; (2) that they were cast into space by lunar volcanoes when the moon was still subject to volcanic action; (3) that they are the products of explosion in the stars. But some astronomers are disposed to think that they originated in a similar manner to other members of the solar system, although it is difficult, on this hypothesis, to account for their great density. The opinion that the iron meteorites have come from the sun, or some other star, is enforced by the fact that they contain hydrogen, carbon, and helium, in forms suggesting that these gases were absorbed while the bodies were immersed in a hot, dense atmosphere.

PART IV. THE FIXED STARS.

PART IV.   THE FIXED STARS.

1. The Stars. The stars are distant suns, varying greatly in remoteness, in magnitude, and in condition. Many of them are much smaller than our sun, and many others are as much larger. They vary, likewise in age, or state of development. Some are relatively young, others in a middle stage, and still others in a condition that may be called solar decrepitude. These proofs of evolution among the stars, the knowledge of which we owe mainly to spectroscopic analysis, serve to establish more firmly the conclusion, to which the simple aspect of the heavens first leads us, that the universe is a connected system, governed everywhere by similar laws and consisting of like materials.

The number of stars visible to the naked eye is about six thousand, but telescopes show tens of millions. It is customary to divide the stars into classes, called magnitudes, according to their apparent brightness. By a system of photometry, or light-measurement, they are grouped into stars of the first, second, third, etc., magnitude. With the naked eye no stars fainter than the sixth magnitude are visible, but very powerful telescopes may show them down to the eighteenth magnitude. Each magnitude is about two and a half times brighter than the next magnitude below in the scale. A first-magnitude star is about one hundred times brighter than one of the sixth magnitude. But, in reality, the variation of brightness is gradual, and for very accurate estimates fractions of a magnitude have to be employed. There are about twenty first-magnitude stars, but they are not all of equal brightness. A more accurate photometry assumes a zero magnitude, very nearly, represented by the star Arcturus, and makes the ratio 2.512. Thus a star, nearly represented by Aldebaran or Altair, which is 2.512 times fainter than the zero magnitude, is of the first magnitude, and a star, nearly represented by the North Star, which is 2.512 times fainter than the first magnitude, is of the second magnitude. Counting in the other direction, a star, like Sirius, which is brighter than the zero magnitude, is said to be of a negative magnitude. The magnitude of Sirius is—1.6. There is only one other star of negative magnitude, Canopus, whose magnitude is—0.9. But for ordinary purposes one need not trouble himself with these refinements.

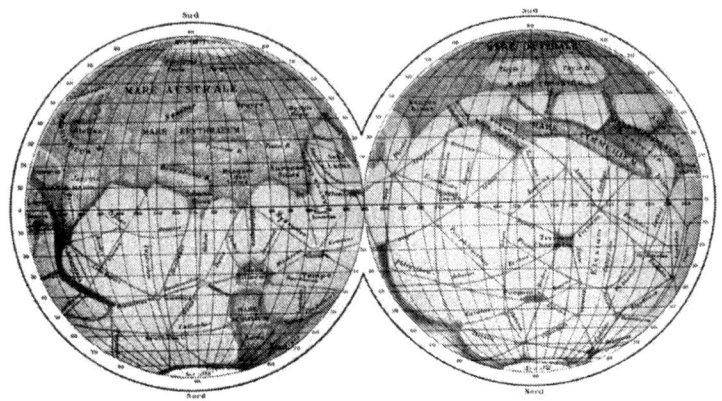

Schiaparelli's Chart of Martian "Canals."

The stars are divided into five principal types, according to their spectra. These are:

I. White stars, having a bluish tinge, in which the spectrum is characterised by broad dark bands, due apparently to an extensive atmosphere of hydrogen, while there are but few lines indicating the presence of metallic vapours. About half the stars whose spectra have been studied belong to Type I.

II. Yellowish-white stars, resembling the sun in having their spectra crossed with a great number of lines produced by metallic vapours, while the hydrogen lines are less conspicuous. These are often called solar stars, and they, too, are very numerous.

III. Orange and slightly reddish stars, whose spectra contain mostly broad bands instead of narrow lines, the bands being situated toward the blue end of the spectrum, whence the prevailing colour, since the blue light is thus cut off. Only a few hundred of these stars are known, but they include most of the well-known variable stars.

IV. Small deep-red stars having dark bands absorbing the light of the red end of the spectrum. Less than a hundred of these stars are known.

V. Stars whose spectra are characterised by bright instead of dark lines, although they also show dark bands. The bright lines indicate that the atmospheric vapours producing them are at a higher temperature than the body of the star. Stars of this type are sometimes called Wolf-Rayet stars and they are few in number.

Various modifications of these main types exist, but we cannot here enter into an account of them. In a general way, although there are exceptions depending upon the precise nature of each spectrum, the white stars are thought to be younger than the yellowish ones, and the red stars older.

In speaking of the "size" of the stars we really mean their luminosity, or the amount of light radiated from them. When a star is said to be a thousand times greater than the sun, the meaning is that the amount of light that it gives would, if both were viewed from the same distance, be equal to a thousand times the amount given by the sun. We have no direct knowledge of the actual size of the stars as globes, because the most powerful telescope is unable to reveal the real disk of a star. In comparing the luminosity of a star with that of the sun its distance must be taken into account. Most of the stars are so far away that we really know nothing of their distances, but there are fifty or more which lie within a distance not too great to enable us to obtain an approximate idea of what it is. The nearest star in the northern sky is so far from being the brightest that it can barely be seen with the naked eye. It must be very much less luminous than the sun. On the other hand, some very bright stars lie at a distance so immense that it can hardly be estimated, and they must exceed the sun in luminosity hundreds and even thousands of times.

The question of the distance of the stars has already been treated in the section on Parallax. In employing our knowledge of star distances for the purpose of comparing their luminosity with that of the sun, we must first ascertain, as accurately as possible, the actual amount of light that the star sends to the earth as compared with the actual amount of light that the sun sends. The star Arcturus gives to our eyes about one forty-billionth as much light as the sun does. Knowing this, we must remember that the intensity of light varies, like gravitation, inversely as the square of the distance. Thus, if the sun were twice as far away as it is, the amount of its light received on the earth would be reduced to one fourth, and if its distance were increased three times, the amount would be reduced to one ninth. If the sun were 200,000 times as far away, its light would be reduced to one forty-billionth, or the same as that of Arcturus. At this point the actual distance of Arcturus enters into the calculation. If that distance were 200,000 times the sun's distance, we should have to conclude that Arcturus was exactly equal to the sun in luminosity, since the sun, if removed to the same distance, would give us the same amount of light. But, in fact, we find that the distance of Arcturus, instead of being 200,000 times that of the sun, is about 10,000,000 times. In other words, it is fifty times as far away as the sun would have to be in order that it should appear to our eyes no brighter than Arcturus. From this it follows that the real luminosity of Arcturus must be the square of 50, or 2500, times that of the sun. In the same manner we find that Sirius, which to the eye appears to be the brightest star in the sky (much brighter than Arcturus because much nearer), is about thirty times as luminous as the sun.

Many of the stars are changeable in brightness, and those in which the changes occur to a notable extent, and periodically, are known as variable stars. It is probable that all the stars, including the sun, are variable to a slight degree. Among the most remarkable variables are Mira, or Omicron

Ceti, in the constellation Cetus, which in the course of about 331 days rises from the ninth to the third magnitude and then falls back again (the maxima of brightness are irregular); and Algol, or Beta Persei, in the constellation Perseus, which, in a period of 2 days, 20 hours, 49 minutes, changes from the third to the second magnitude and back again. In the case of Mira the cause of the changes is believed to lie in the star itself, and they may be connected with its gradual extinction. The majority of the variable stars belong to this class. As to Algol, the variability is apparently due to a huge dark body circling close around the star with great speed, and periodically producing partial eclipses of its light. There are a few other stars with short periods of variability which belong to the class of Algol.

When examined with telescopes many of the stars are found to be double, triple, or multiple. Often this arises simply from the fact that two or more happen to lie in nearly the same line of sight from the earth, but in many other cases it is found that there is a real connection, and that the stars concerned revolve, under the influence of their mutual gravitation, round a common centre of force. When two stars are thus connected they are called a binary. The periods of revolution range from fifty to several hundred years. Among the most celebrated binary stars are Alpha Centauri, in the southern hemisphere, the nearest known star to the solar system, whose components revolve in a period of about eighty years; Gamma Virginis, in the constellation Virgo, period about one hundred and seventy years; and Sirius, period about fifty-three years. In the case of Sirius, one of the components, although perhaps half as massive as its companion, is ten thousand times less bright.

There is another class of binary stars, in which one of the companions is invisible, its presence being indicated by the effects of its gravitational pull upon the other. Algol may be regarded as an example of this kind of stellar association. But there are stars of this class, where the companion causes no eclipses, either because it is not dark, or because it never passes over the other, as seen from the earth, but where its existence is proved, in a very interesting way, by the spectroscope. In these stars, called spectroscopic binaries, two bright components are so close together that no telescope is able to make them separately visible, but when their plane of revolution lies nearly in our line of sight the lines in their combined spectrum are seen periodically split asunder. To understand this, we must recall the principles underlying spectroscopic analysis and add something to what was said before on that subject.

Light consists of waves in the ether of different lengths and making upon the eye different impressions of colour according to the length of the waves. The longest waves are at the red end of the spectrum and the shortest at the blue, or violet, end. But since they all move onward with the same speed, it is clear that the short blue waves must fall in quicker succession on the retina of the eye than the long red waves. Now suppose that the source of light from which the waves come is approaching very swiftly; it is easy to see that all the waves will strike the eye with greater rapidity, and that the whole spectrum will be shifted toward the blue, or short-wave, end. The Fraunhofer lines will share in this shifting of position. Next suppose that the source of the light is retreating from the eye. The same effect will occur in a reversed sense, for now there will be a general shift toward the red end of the spectrum. A sufficiently clear illustration, by analogy, is furnished by the waves of sound. We know that low-pitched sounds are produced by long waves, and high-pitched ones by short waves; then if the source of the sound, such as a locomotive whistle, rapidly approaches the ear the waves are crowded together, or shifted as a whole toward the short end of the gamut, whereupon the sound rises to a shrill scream. If, on the contrary, the source of sound is retreating, the shift is in the other direction, and the sound drops to a lower pitch.

This is precisely what happens in the spectrum of a star which is either approaching or receding from the eye. If it is approaching, the Fraunhofer lines are seen shifted out of their normal position toward the blue, and if it is receding they are shifted toward the red. The amount of shifting will depend upon the speed of the star's motion. If that motion is across the line of sight there will be no shifting, because then the source of light is neither approaching nor receding. Now take the case of a binary star whose components are too close to be separated by a telescope. If they happen to be revolving round their common centre in a plane nearly coinciding with the line of sight from the earth, one of them must be approaching the eye at the same time that the other is receding from it, and the consequence is that the spectral lines of the first will be shifted toward the blue, while those of the second are shifted toward the red. The colours of the two intermingled spectra blend into each other too gradually to enable this effect to be detected by their means, but the Fraunhofer lines are sharply defined, and in them the shift is clearly seen; and since there is a simultaneous shifting in opposite directions the lines appear split. But when the two stars are in that part of their orbit where their common motion is across the line of sight the lines close up again, because then there is no shift. This phenomenon is beautifully exhibited by one of the first spectroscopic binaries to be discovered, Beta Aurigæ. In 1889, Prof. E. C. Pickering noticed that the spectral lines of this star appeared split every second night, from which he inferred that it consisted of two stars revolving round a common centre in a period of four days.

This spectroscopic method has been applied to determine the speed with which certain single stars are approaching or receding from the solar system. It has also served to show, what we have before remarked, that the inner parts of Saturn's rings travel faster than the outer parts. Moreover, it has been used in measuring the rate of the sun's rotation on its axis, for it is plain that one edge of the sun approaches us while the opposite edge is receding. Even the effect of the rotation of Jupiter has been revealed in this way, and the same method will probably settle the question whether Venus rotates rapidly, or keeps the same face always toward the sun.

Not only do many stars revolve in orbits about near-by companions, but all the stars, without exception, are independently in motion. They appear to be travelling through space in many different directions, each following its own chosen way without regard to the others, and each moving at its own gait. These movements of the stars are called proper motions. The direction of the sun's proper motion is, roughly speaking, northward, and it travels at the rate of twelve or fourteen miles per second, carrying the earth and the other planets along with it. Some stars have a much greater speed than the sun, and some a less speed. As we have said, these motions are in many different directions, and no attempt to discover any common law underlying them has been entirely successful, although it has been found that in some parts of the sky a certain number of stars appear to be travelling along nearly parallel paths, like flocks of migrating birds. In recent years some indications have been found of the possible existence of two great general currents of movement, almost directly opposed to each other, part of the stars following one current and part the other. But no indication has been discovered of the existence of any common centre of motion. Several relatively near-by stars appear to be moving in the same direction as the sun. Stars that are closely grouped together, like the cluster of the Pleiades, seem to share a common motion of translation through space. We have already remarked that when stars are found to be moving toward or away from the sun, spectroscopic observation of the shifting of their lines gives a means of calculating their velocity. In other cases, the velocity across the line of sight can be calculated if we know the distance of the stars concerned. One interesting result of the fact that the earth goes along with the sun in its flight is that the orbit of the earth cannot be a closed curve, but must have the form of a spiral in space. In consequence of this we are continually advancing, at the rate of at least 400,000,000 miles per year, toward the northern quarter of the sky. The path pursued by the sun appears to be straight, although it may, in fact, be a curve so large that we are unable in the course of a lifetime, or many lifetimes, to detect its departure from a direct line. At any rate we know that, as the earth accompanies the sun, we are continually moving into new regions of space.

It has been stated that many millions of stars are visible with telescopes—perhaps a hundred millions, or even more. The great majority of these are found in a broad irregular band, extending entirely round the sky, and called the Milky Way, or the Galaxy. To the naked eye the Milky Way appears as a softly shining baldric encircling the heavens, but the telescope shows that it consists of multitudes of faint stars, whose minuteness is probably mainly due to the immensity of their distance, although it may be partly a result of their relative lack of actual size, or luminosity. In many parts of the Milky Way the stars appear so crowded that they present the appearance of sparkling clouds. The photographs of these aggregations of stars in the Milky Way, made by Barnard, are marvellous beyond description. In the Milky Way, and sometimes outside it, there exist globular star-clusters, in which the stars seem so crowded toward the centre that it is impossible to separate them with a telescope, and the effect is that of a glistering ball made up of thousands of silvery particles, like a heap of microscopic thermometer bulbs in the sunshine. A famous cluster of this kind is found in the constellation Hercules.

The Milky Way evidently has the form of a vast wreath, made up of many interlaced branches, some of which extend considerably beyond its mean borders. Within, this starry wreath space is relatively empty of stars, although some thousands do exist there, of which the sun is one. We are at present situated not very far from the centre of the opening within the ring or wreath, but the proper motion of the sun is carrying us across this comparatively open space, and in the course of time, if the direction of our motion does not change, we shall arrive at a point not far from its northern border. The Milky Way probably indicates the general plan on which the visible universe is constructed, or what has been called the architecture of the heavens, but we still know too little of this plan to be able to say exactly what it is.

The number of stars in existence at any time varies to a slight degree, for occasionally a star disappears, or a new one makes its appearance. These, however, are rare phenomena, and new stars usually disappear or fade away after a short time, for which reason they are often called temporary stars. The greatest of these phenomena ever beheld was Tycho Brahe's star, which suddenly burst into view in the constellation Cassiopeia in 1572, and disappeared after a couple of years, although at first it was the brightest star in the heavens. Another temporary star, nearly as brilliant, appeared in the constellation Perseus, in 1901, and this, as it faded, gradually turned into a nebula, or a star surrounded by a nebula. It is generally thought that outbursts of this kind are caused by the collision of two or more massive bodies, which were invisible before their disastrous encounter in space. The heat developed by such a collision would be sufficient to vaporise them, and thus to produce the appearance of a new blazing star. It is possible that space contains an enormous number of great obscure bodies,—extinguished suns, perhaps—which are moving in all directions as rapidly as the visible stars.

2. The Nebulæ. These objects, which get their name from their cloud-like appearance, are among the most puzzling phenomena of the heavens, although they seem to suggest a means of explaining the origin of stars. Many thousands of nebulæ are known, but there are only two or three bright enough to be visible to the naked eye. One of these is in the "sword" of the imaginary giant figure marking the constellation Orion, and another is in the constellation Andromeda. They look to the unaided eye like misty specks, and require considerable attention to be seen at all. But in telescopes their appearance is marvellous. The Orion nebula is a broad, irregular cloud, with many brighter points, and a considerable number of stars intermingled with it, while the Andromeda nebula has a long spindle shape, with a brighter spot in the centre. It is covered and surrounded with multitudes of faint stars. It was only after astronomical photography had been perfected that the real shapes of the nebulæ were clearly revealed. Thousands of nebulæ have been discovered by photography, which are barely if at all visible to the eye, even when aided by powerful telescopes. This arises from the fact that the sensitive photographic plate accumulates the impression that the light makes upon it, showing more and more the longer it is exposed. Plates placed in the focus of telescopes, arranged to utilise specially the "photographic rays," are often exposed for many hours on end in order to picture faint nebulæ and faint stars, so that they reveal things that the eye, which sees all it can see at a glance, is unable to perceive.

Nebulæ are generally divided into two classes—the "white" nebulæ and the "green" nebulæ. The first, of which the Andromeda nebula is a striking example, give a continuous spectrum without dark lines, as if they consisted either of gas under high pressure, or of something in a solid or liquid state. The second, conspicuously represented by the Orion nebula, give a spectrum consisting of a few bright lines, characteristic of such gases as hydrogen and helium, together with other substances not yet recognised. But there is no continuous spectrum like that shown by the white nebulæ, from which it is inferred that the green nebulæ, at least, are wholly gaseous in their constitution. The precise constitution of the white nebulæ remains to be determined.

It is only in relatively recent years that the fact has become known that the majority of nebulæ have a spiral form. There is almost invariably a central condensed mass from which great spiral arms wind away on all sides, giving to many of them the appearance of spinning pin-wheels, flinging off streams of fire and sparks on all sides. The spirals look as if they were gaseous, but along and in them are arrayed many condensed knots, and frequently curving rows of faint stars are seen apparently in continuation of the nebulous spirals. The suggestion conveyed is that the stars have been formed by condensation from the spirals. These nebulæ generally give the spectra of the white class, but there are also sometimes seen bright lines due to glowing gases. The Andromeda nebula is sometimes described as spiral, but its aspect is rather that of a great central mass surrounded with immense elliptical rings, some of which have broken up and are condensing into separate masses. The Orion nebula is a chaotic cloud, filled with partial vacancies and ribbed with many curving, wave-like forms.

There are other nebulæ which have the form of elliptical rings, occasionally with one or more stars near the centre. A famous example of this kind is found in the constellation Lyra. Still others have been compared in shape to the planet Saturn with its rings, and some are altogether bizarre in form, occasionally looking like glowing tresses floating among the stars.

The apparent association of nebulæ with stars led to the so-called nebular hypothesis, according to which stars are formed, as already suggested, by the condensation of nebulous matter. In the celebrated form which Laplace gave to this hypothesis, it was concerned specially with the origin of our solar system. He assumed that the sun was once enormously expanded, in a nebulous state, or surrounded with a nebulous cloud, and that as it contracted rings were left off around the periphery of the vast rotating mass. These rings subsequently breaking and condensing into globes, were supposed to have given rise to the planets. It is still believed that the sun and the other stars may have originated from the condensation of nebulæ, but many objections have been found to the form in which Laplace put his hypothesis, and the discovery of the spiral nebulæ has led to other conjectures concerning the way in which the transformation is brought about. But we have not here the space to enter into this discussion, although it is of fascinating interest.

A word more should be said about the use of photography in astronomy. It is hardly going too far to aver that the photographic plate has taken the place of the human retina in recording celestial phenomena, especially among the stars and nebulæ. Not only are the forms of such objects now exclusively recorded by photography, but the spectra of all kinds of celestial objects— sun, stars, nebulæ, etc.—are photographed and afterward studied at leisure. In this way many of the most important discoveries of recent years have been made, including those of variable stars and new stars. Photographic charts of the heavens exist, and by comparing these with others made later, changes which would escape the eye can be detected. Comets are sometimes, and new asteroids almost invariably, discovered by photography. The changes in the spectra of comets and new stars are thus recorded with an accuracy that would be otherwise unattainable. Photographs of the moon excel in accuracy all that can be done by manual drawing, and while photographs of the planets still fail to show many of the fine details visible with telescopes, continual improvements are being made. Many of the great telescopes now in use or in course of construction are intended specially for photographic work.

3. The Constellations. The division of the stars into constellations constitutes the uranography or the "geography of the heavens." The majority of the constellations are very ancient, and their precise origin is unknown, but those which are invisible from the northern hemisphere have all been named since the great exploring expeditions to the south seas. There are more than sixty constellations now generally recognised. Twelve of these belong to the zodiac, and bear the same names as the zodiacal signs, although the precession of the equinoxes has drifted them out of their original relation to the signs. Many of the constellations are memorials of prehistoric myths, and a large number are connected with the story of the Argonautic expedition and with other famous Greek legends. Thus the constellations form a pictorial scroll of legendary history and mythology, and possess a deep interest independent of the science of astronomy. For their history and for the legends connected with them, the reader who desires a not too detailed résumé, may consult *Astronomy with the Naked Eye*, and for guidance in finding the constellations, *Astronomy with an Opera-glass*, or *Roundthe Year with the Stars*. The quickest

way to learn the constellations is to engage the aid of some one who knows
them already, and can point them out in the sky. The next best way is to use
star charts, or a star-finder or planisphere.

A considerable number of the brighter and more important stars are
known by individual names, such as Sirius, Canopus, Achernar, Arcturus,
Vega, Rigel, Betelgeuse, Procyon, Spica, Aldebaran, Regulus, Altair, and
Fomalhaut.   Astronomers usually designate the principal stars of each
constellation by the letters of the Greek alphabet, , , , etc., the brightest
star in the constellation bearing the name of the first letter, the next brightest
that of the second letter, and so on.

The constellations are very irregular in outline, and their borders are only
fixed with sufficient definiteness to avoid the inclusion of stars catalogued as
belonging to one, within the limits of another. In all cases the names come
from some fancied resemblance of the figures formed by the principal stars of
the constellation to a man, woman, animal, or other object. In only a few
cases are these resemblances very striking.

The most useful constellations for the beginner are those surrounding
the north celestial pole, and we give a little circular chart showing their
characteristic stars.   The names of the months running round the circle
indicate the times of the year when these constellations are to be seen on
or near the meridian in the north. Turn the chart so that the particular
month is at the bottom, and suppose yourself to be facing northward. The
hour when the observation is supposed to be made is, in every case, about 9
o'clock in the evening, and the date is about the first of the month. The top
of the chart represents the sky a little below the zenith in the north, and the
bottom represents the horizon in the north.

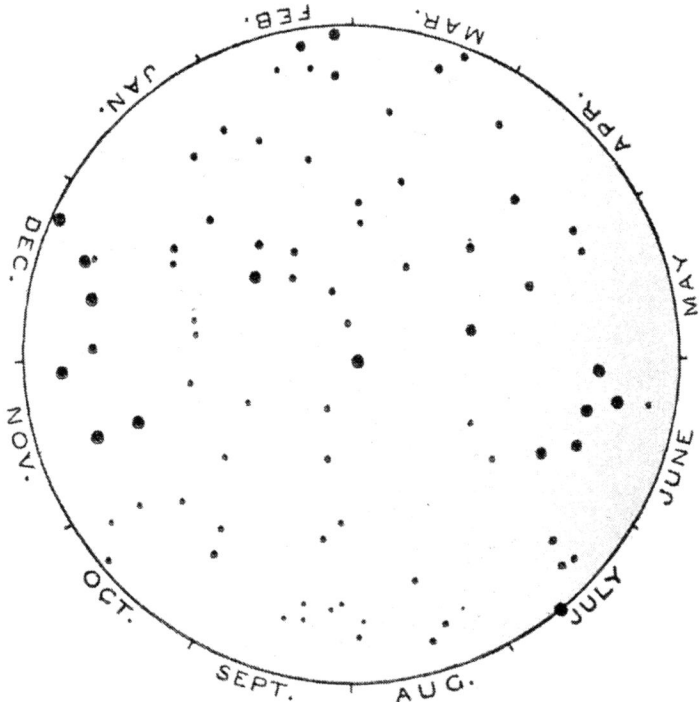

Fig. 18. The North Circumpolar Stars.

The apparent yearly revolution of the heavens, resulting from the motion of the earth in its orbit, causes the constellations to move westward in a circle round the pole, at the rate of about 30° per month. But the daily rotation of the earth on its axis causes a similar westward motion of the heavens, at the rate of about 30° for every two hours. From this it results that on the same night, after an interval of two hours, you will see the constellations occupying the place that they will have, at the original hour of observation, one month later. Thus, if you observe their positions at 9 P.M. on the first of January, and then turn the chart so as to bring February at the bottom, you will see the constellations around the north pole of the heavens placed as they will be at 11 P.M. on the first of January.

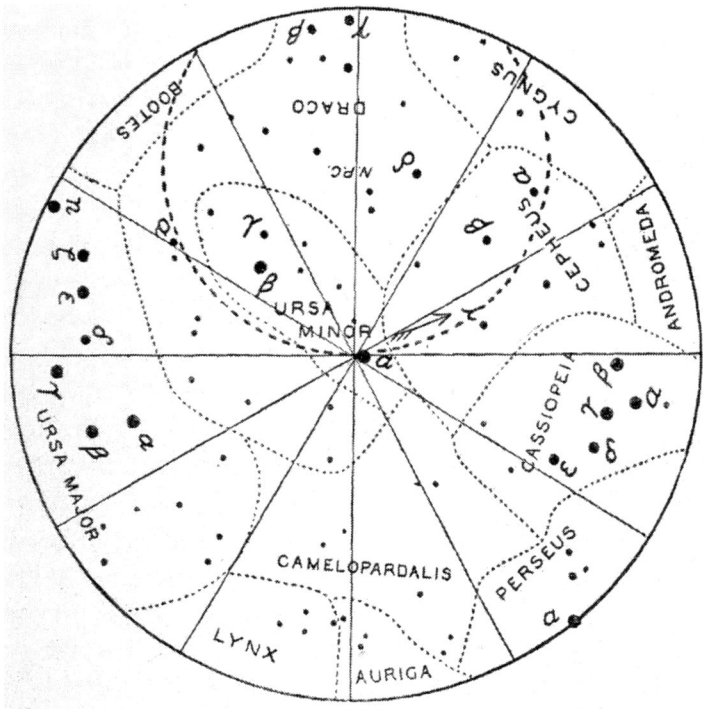

Fig. 19. Key to North Circumpolar Stars.

Only the conspicuous stars have been represented in the chart, just enough being included to enable the learner to recognise the constellations by their characteristic star groups, from which they have received their names. The chart extends to a distance of 40° from the pole, so that, for observers situated in the mean latitude of the United States, none of the constellations represented ever descends below the horizon, those that are at the border of the chart just skimming the horizon when they are below the pole.

On the key to the chart the Greek-letter names of the principal stars have been attached, but some of them have other names which are more picturesque. These are as follows: In Ursa Major (the Great Bear, which includes the Great Dipper), is called Dubhe,   Merak,   Phaed,   Megrez, Alioth,   Mizar, and   Benetnash. The little star close by Mizar is Alcor. In Cassiopeia, is called Schedar,   Caph, and   Ruchbar. In Ursa Minor, the Little Bear, is called Polaris, or the North Star, and   Kochab. In Draco, is called Thuban, and   Eltanin. In Cepheus, is called Alderamin, and Alfirk. These names are nearly all of Arabic origin. It will be observed that Merak and Dubhe are the famous "Pointers," which serve to indicate the position of the North Star, while Thuban is the "star of the pyramid," before mentioned. The north celestial pole is situated almost exactly on a straight line drawn from Mizar through the North Star to Ruchbar, and a little more than a degree from the North Star in the direction of Ruchbar. This furnishes a ready means for ascertaining the position of the meridian. For instance, about the middle of October, Mizar is very close to the meridian below the pole, and Ruchbar equally close to it above the pole, and then, since the North Star is in line with these two, it also must be practically on the meridian, and its direction indicates very nearly true north. The same method is applicable whenever, at any other time of the year or of the night, Mizar and Ruchbar are observed to lie upon a vertical line, no matter which is above and which below. It is also possible to make a very good guess at the time of night by knowing the varying position of the line joining these stars.

The star Caph is an important landmark because it lies almost on the great circle of the equinoctial colure, which passes through the vernal and autumnal equinoxes.

On the key, the location of the North Pole of the Ecliptic is shown, and the greater part of the circle described by the north celestial pole in the period of 25,800 years.

While the reader who wishes to pursue the study of the constellations in detail must be referred to some of the works before mentioned, or others of like character, it is possible here to aid him in making a preliminary acquaintance with other constellations beside those included in our little chart, by taking each of the months in turn, and describing the constellations which he will see on or near the meridian south of the border of the chart at the same time that the polar constellations corresponding to the month selected are on or near the meridian in the north. Thus, at 9 P.M. about the first of January, the constellation Perseus, lying in a rich part of the Milky Way, is nearly overhead and directly south of the North Star. This constellation is marked by a curved row of stars, the brightest of which, of the second magnitude, is Algenib, or Persei. A few degrees south-west of Algenib is the wonderful variable Algol. East of Perseus is seen the very brilliant white star Capella in the constellation Auriga. This is one of the brightest stars in the sky. Almost directly south of Perseus, the eye will be caught by the glimmering cluster of the Pleiades in the constellation Taurus. A short distance south-east of the Pleiades is the group of the Hyades in Taurus, shaped like the letter V, with the beautiful reddish star Aldebaran in the upper end of the southern branch of the letter. The ecliptic runs between the Pleiades and the Hyades. Still lower in the south will be seen a part of the long-winding constellation Eridanus, the River Po. Its stars are not bright but they appear in significant rows and streams.

About the first of February the constellation Auriga is on the meridian not far from overhead, Capella lying toward the west. Directly under Auriga, two rather conspicuous stars mark the tips of the horns of Taurus, imagined as a gigantic bull, and south of these, with its centre on the equator, scintillates the magnificent constellation Orion, the most splendid in all the sky, with two great first-magnitude stars, one, in the shoulder of the imaginary giant, of an orange hue, called Betelguese, and the other in the foot, of a blue-white radiance, called Rigel. Between these is stretched the straight line of the "belt," consisting of three beautiful second-magnitude stars, about a degree and a half apart. Their names, beginning with the western one, are Mintaka, Alnilam, and Alnitah. Directly under the belt, in the midst of a short row of faint stars called the "sword," is the great Orion nebula. It will be observed that the three stars of the belt point, though not exactly, toward the brightest of all stars, Sirius, in the constellation Canis Major, the Great Dog, which is seen advancing from the east. Under Orion is a little constellation named Lepus, the Hare.

The first of March the region overhead is occupied by the very faint constellation Lynx.   South of it, and astride the ecliptic, appear the constellations Gemini, the Twins, and Cancer, the Crab. These, like Taurus, belong to the zodiac. The Twins are westward from Cancer, and are marked by two nearly equal stars, about five degrees apart. The more westerly and northerly one is Castor and the other is Pollux. Cancer is marked by a small cluster of faint stars called Præsepe, the Manger (also sometimes the Beehive). Directly south of the Twins, is the bright lone star Procyon, in the constellation Canis Minor, the Little Dog. Sirius and the other stars of Canis Major, which make a striking figure, are seen south-west of Procyon.

The first of April the zodiac constellation Leo is near the meridian, recognisable by a sickle-shaped figure marking the head and breast of the imaginary Lion. The bright star at the end of the handle of the sickle is Regulus. Above Leo, between it and the Great Dipper, appears a group of stars belonging to the small constellation Leo Minor, the Little Lion. Farther south is a winding ribbon of stars indicating the constellation Hydra, the Water Serpent. Its chief star, Alphard, of a slightly reddish tint, is seen west of the meridian and a few degrees south of the equator.

At the beginning of May, when the Great Dipper is nearly overhead, the small constellation Canes Venatici, the Hunting Dogs, is seen directly under the handle of the Dipper, and south of that a cobwebby spot, consisting of minute stars, indicates the position of the constellation Coma Berenices, Berenice's Hair. Still farther south, where the ecliptic and the equator cross, at the autumnal equinox, is the large constellation Virgo, the Virgin, also one of the zodiacal band. Its chief star Spica, a pure white gem, is seen some 20° east of the meridian. Below and westward from Virgo, and south of the equator, are the constellations Crater, the Cup, and Corvus, the Crow. The stars of Hydra continue to run eastward below these constellations. The westernmost, Crater, consists of small stars forming a rude semicircle open toward the east, while Corvus, which possesses brighter stars, has the form of a quadrilateral.

The first of June the great golden star Arcturus, whose position may be found by running the eye along the curve of the handle of the Great Dipper, and continuing onward a distance equal to the whole length of the Dipper, is

seen approaching the meridian from the east and high overhead. This superb
star is the leader of the constellation Boötes, the Bear-Driver. Spica in Virgo
is now a little west of the meridian.

The first of July, when the centre of Draco is on the meridian north of the
zenith, the exquisite circlet of stars called Corona Borealis, the Northern
Crown, is nearly overhead. A short distance north-east of it appears a
double-quadrilateral figure, marking out the constellation Hercules, while
directly south of the Crown a crooked line of stars trending eastward indicates
the constellation Serpens, the Serpent. South-west of Serpens, two widely
separated but nearly equal stars of the second magnitude distinguish the
zodiacal constellation Libra, the Balance; while lower down toward the south-
east appears the brilliant red star Antares, in the constellation Scorpio,
likewise belonging to the zodiac.

On the first of August the head of Draco is on the meridian near the
zenith, and south of it is seen Hercules, toward the west, and the exceedingly
brilliant star Vega, in the constellation Lyra, the Lyre, toward the east. Vega,
or Alpha Lyræ, has few rivals for beauty. Its light has a decided bluish-white
tone, which is greatly accentuated when it is viewed with a telescope. South
of Hercules two or three rows of rather large, widely separated stars mark
the constellation Ophiuchus, the Serpent-Bearer. This extends across the
equator. Below it, in a rich part of the Milky Way, is Scorpio, whose winding
line, beginning with Antares west of the meridian, terminates a considerable
distance east of the meridian in a pair of stars representing the uplifted sting
of the imaginary monster.

The first of September the Milky Way runs directly overhead, and in the
midst of it shines the large and striking figure called the Northern Cross, in
the constellation Cygnus, the Swan. The bright star at the head of the Cross
is named Denib. Below the Cross and in the eastern edge of the Milky Way
is the constellation Aquila, the Eagle, marked by a bright star, Altair, with
a smaller one on each side and not far away. Low in the south, a little west
of the meridian and partly immersed in the brightest portion of the Milky
Way, is the zodiacal constellation Sagittarius, the Archer. It is distinguished
by a group of stars several of which form the figure of the upturned bowl of
a dipper, sometimes called the Milk Dipper. East of Cygnus and Aquila a
diamond-shaped figure marks the small constellation Delphinus, the Dolphin.

At the opening of October, when Denib is near the meridian, the sky
directly in the south is not very brilliant. Low down, south of the equator, is
seen the zodiacal constellation Capricornus, the Goat, with a noticeable pair
of stars in the head of the imaginary animal.

On the first of November, when Cassiopeia is approaching the meridian
overhead, the Great Square, in the constellation Pegasus, is on the meridian
south of the zenith, while south-west of Pegasus the zodiacal constellation
Aquarius, the Water-Bearer, appears on the ecliptic. A curious scrawling Y-
shaped figure in the upper part of Aquarius serves as a mark to identify the

constellation. Thirty degrees south of this shines the bright star Fomalhaut, in the constellation Piscis Australis, the Southern Fish. The two stars forming the eastern side of the Great Square of Pegasus are interesting because, like Caph in Cassiopeia, they lie close to the line of the equinoctial colure. The northern one is called Alpheratz and the southern Gamma Pegasi. Alpheratz is a star claimed by two constellations, since it not only marks one corner of the square of Pegasus, but it also serves to indicate the head of the maiden in the celebrated constellation of Andromeda.

The first of December, Andromeda is seen nearly overhead, south of Cassiopeia. The constellation is marked by a row of three second-magnitude stars, beginning on the east with Alpheratz and terminating near Perseus with Almaack. The central star is named Mirach. A few degrees north-west of Mirach glimmers the great Andromeda nebula. Below Andromeda, west of the meridian, appears the zodiacal constellation Aries, the Ram, indicated by a group of three stars, forming a triangle, the brightest of which is called Hamal. South-westerly from Aries is the zodiacal constellation Pisces, the Fishes, which consists mainly of faint stars arranged in pairs and running far toward the west along the course of the ecliptic, which crosses the equator at the vernal equinox, near the western end of the constellation. South of Pisces and Aries is the broad constellation Cetus, the Whale, marked by a number of large quadrilateral and pentagonal figures, formed by its stars. Near the centre of this constellation, but not ordinarily visible to the naked eye, is the celebrated variable Mira, also known as Omicron Ceti.

With a little application any person can learn to recognise these constellations, even with the slight aid here offered, and if he does, he will find the knowledge thus acquired as delightful as it is useful.

INDEX

www.ingramcontent.com/pod-product-compliance
Ingram Content Group UK Ltd.
Pitfield, Milton Keynes, MK11 3LW, UK
UKHW021426020126
9874UKWH00026B/373